COLLIERY SETTLEMENT IN THE SOUTH WALES

COALFIELD, 1850 TO 1926

D1388719

t in the South Wales

University of Hull Publications

Occasional Papers in Geography No. 14

General Editor: H. R. Wilkinson

*Professor of Geography in the*
*University of Hull*

# COLLIERY SETTLEMENT IN THE
## SOUTH WALES COALFIELD
### 1850–1926

PHILIP N. JONES

University of Hull Publications
1969

Made and Printed in England
by Hull Printers Limited
Great Gutter Lane, Willerby, Hull, HU10 6DH

# CONTENTS

*v*

University of Hull Publications
General Editor: H. R. Wilkinson
*Professor of Geography in the
University of Hull*
The University of Hull 1969

# FIGURES

# PLATES

# ACKNOWLEDGEMENTS

The author wishes to thank Mr. M. B. Stedman and Professor D. L. Linton for their interest in his work on mining settlement and encouragement when he was a research student at Birmingham. He is grateful to the staffs of the National Library of Wales, Aberystwyth, the Glamorgan County Record Office, Cardiff, the British Transport Commission Archives, London, for their efficiency in providing the bulk of the documentary sources used in the research. He is indebted also to local authorities and many individual officers of Glyncorrwg U.D., Llantrisant and Llantwit Fardre R.D., Maesteg U.D., Neath R.D., Ogmore and Garw U.D. and Rhondda U.D. for their co-operation in making available their registers of deposited plans and samples of the actual deposited plans for him to study; whilst their hospitality in making room available for many consecutive days of study is warmly appreciated. He would like to thank Mr. R. Dean, Chief Technician of the Department of Geography, University of Hull, for his advice on the preparation of the final maps and diagrams, which have been so ably executed by the technicians in the Drawing Office, Miss S. Hakeney, Miss S. Hillman, and Mr. K. Scurr. Finally, a very considerable debt is owed to Professor H. R. Wilkinson of the Department of Geography at Hull for the advice and criticism which he has given both in organizing the final form and structure of the paper in particular, and on many aspects of detailed presentation.

PHILIP N. JONES

CHAPTER I

# THE SIGNIFICANCE OF COLLIERY SETTLEMENT AS A FIELD FOR RESEARCH

*Introduction*

COLLIERY settlement, and indeed the broader topic of industrial settlement, has been a neglected field of geographical enquiry and publication in English-language journals, suggesting a lack of interest in this important settlement form. This is particularly unfortunate in countries such as the United Kingdom and the United States where industrial settlement was of great importance in the formative period of industrialization in the nineteenth and early twentieth centuries, and where the legacy is still very much in evidence today, both in the landscape and at the core of the social geography of many industrial areas. The lack of research forms a notable contrast to the acknowledged contributions which geographers have made, and are making, to the understanding of urban and rural settlement.

Most studies of coalfield regions, particularly in Britain, have been more concerned with the economic adjustments made necessary by the contraction of the coal industry since the 1920s, and show little interest in the particular features of settlement structure within the various regions (e.g., Daysh, *et al.*, Gendarme). Apart from the pioneering work of Smailes we have little in the way of a comprehensive study of the development, structure and morphology of colliery settlement, worthy in itself of detailed analysis (Smailes). The lack of appreciation of the complexity of settlement in coalfield regions is apparent, and has created a gap in geographical analysis and concepts concerning the interpretation and significance of mining settlement. This has inevitably led to the errors of over-simplification and misinterpretation, and it is hoped that this Paper will present the field of colliery settlement in a more adequate perspective. In particular, very little consideration has been given to the quite elementary and obvious point that whilst settlement of any type is a very permanent and 'fixed' element in locational terms, the collieries are by their very nature exhaustive, and therefore to a greater or lesser degree 'mobile' elements, in that patterns of exploitation have continually fluctuated in any coalfield in response to the exigencies of market and production factors. This basic paradox poses innumerable problems concerning the relationship between static and dynamic elements in the creation of an overall pattern of settlement. Nevertheless the striking individuality of the settlement form has rarely been questioned, and it is clear that colliery districts are distinguished by a certain homogeneity of settlement form and demographic structure. These two factors are constant themes underlying all geographical work in regions of colliery settlement. Most geographers have also emphasised the incomplete socio-economic structure of colliery settlements, which is the main barrier to the achievement of full urban status. One of the distinguishing aspects, therefore, is the elementary form of the individual colliery settlement, most being initially little more than adjuncts to an economic enterprise, the colliery, and having little relationship to the hinterland.

Sorre and Schwarz have contributed significantly to the body of generalizations which now exist for colliery settlement as a distinctive form, and which have thus become a foundation for subsequent work. Sorre particularly stressed the necessity for examining closely the relationship of mining settlement to the pre-existing rural pattern (Sorre). Thus we have a generalized but nevertheless critical distinction between colliery settlement established in a previously well-settled area, and colliery settlement which was essentially 'pioneering' in relation to its environment. The second generalization concerns the nature of the basic unit of exploitation, particularly its employment capacity, which obviously was of major importance as a determinant of settlement response. This factor assumes added significance since technological progress after 1800 led to a continual increase in the scale of individual mining units. Schwarz has emphasised the need to take this time-factor into account, and Smailes has shown its relevance to the understanding of colliery settlement in the north-east England coalfield (Schwarz, Smailes).

*Sorre's typology*

Sorre recognized three basic types of colliery settlement, based on an amalgamation of developmental features and scale of mining activity, and these categories are outlined here as forming the basis of most generalizations about colliery settlement to date. Where colliery settlement has developed in a formerly well-settled area, two processes are involved. New settlement units can be established, in varying degrees of locational proximity to the existing settlement; and existing settlement can be enlarged in a rather haphazard and untidy fashion by new units which nevertheless are an organic part of the whole. Both types can occur in the same coalfield, and the prevalence of one or the other depends on a whole set of economic, cultural and social factors.

The first category has been described by Sorre as 'the large compact mining settlement of stereotyped plan with single-family dwellings'. The type occurs mainly on coalfields which were previously well-settled, but the key factor is that the associated units of exploitation were large or very large. From the mid-nineteenth century to the Great War these settlement units were composed of long streets of terraced houses usually arranged in rigid grid-plan, giving rise to the 'corons' of France, the 'pit-villages' of northern England, and the 'Reihenhaus-Siedlungen' of the German coalfields. After 1918 changed social conditions saw the introduction of 'cottage-style' dwellings with gardens, arranged in plans geometrically more varied than the grid, but in all other respects just as monotonous and stereotyped. But the fundamental purpose of the settlement unit did not change—to provide living accommodation as close as possible to a particular colliery, and little thought was given to the social desirability of a co-ordinated programme of urban development in any European coalfield. This type of settlement unit has graphically been termed the 'cité minière' by the French; in other European countries its distinctiveness has resulted in terms such as 'zwijnwijk' in the Flemish-speaking Kempenland coalfield, 'bergarbeiterkolonien' in Germany, and 'garden-village layout' in Britain.

The second category occurs when colliery settlement forms what Sorre has termed a 'nebulous accretion' on the rural, agricultural stratum, resulting in a linear expansion of existing nuclei along major and minor thoroughfares, with thickenings of settlement at certain points. This has occurred on a large scale in the Borinage section of the south Belgium coalfield (Lefévre, Demangeon), but the basic type also occurs in many British coalfields, particularly sections of the west Midlands coalfields, or the Carmarthenshire part of the anthracite coalfield of south Wales.

The third category is the pattern of settlement dominated by isolated mining hamlets, which finds its best expression in the Appalachian coalfields of West Virginia, Virginia, Kentucky and Tennessee (Murphy). In this region colliery settlement is a pioneer element in that the rural stratum was very scanty and dispersed. The small size of the collieries and associated settlements, their cramped valley-bound sites, and the sheer isolation in a previous wilderness area led to the establishment of a pattern of mining communities unequalled for their primitiveness and poverty of social facilities (Baulig, Gottman). This scheme, and other related attempts at generalization, are obviously only an introduction to the study of the problems surrounding colliery settlement. Characteristic of all the schemes is the narrowness of conception of the form of colliery settlement, which is almost always seen as a kind of symbiosis of colliery company and workforce, when in reality most situations displayed a greater complexity resulting from combinations of historical, cultural and economic factors which can only be understood in terms of each coalfield.

## RECURRENT ELEMENTS IN THE COLLIERY SETTLEMENT COMPLEX

In most coalfields for which geographical accounts exist, mining settlement forms part of a visible landscape association, a formal complex, in which housing, collieries and associated works, railways and mineral lines, spoil tips, coal company offices, and mine manager's residence are invariably present. In larger settlements other buildings exist to serve elementary social needs, such as churches, schools, cafes or inns, workmen's institutes, and shops, but obviously much depends on the population involved, amongst other factors. In the case of the Virginian coalfield, Gottman states that schools were normally found in mining villages only where the coal companies were of a benevolent disposition (Gottman). The very distinctive combination of landscape elements, which convey a strong character to a mining region, are matched by the no less significant existence of strong community ties which are not born out of deep roots, or population stability, but out of the demands of a rigorous and dangerous occupation.

And yet, in an analysis of colliery settlement and its structural patterns, it is more important to understand the total functional complex, of which actual colliery *settlement* formed only one part. In many respects, the main elements of the functional complex of mining recur throughout Europe and North America. They were, in broad outline, orientated absolutely to the fulfilment of one economic role—to exploit coal profitably. This meant, for most companies in most coalfields, leasing a concession, finding and raising coal, and distributing and marketing that coal at a price which yielded a net return on capital employed. All other elements dovetail into this 'critical path'—the selling price of competing companies with better or poorer concessions, the creation of new markets for different coal types, the provision of communications to ship coal outwards, the gathering together of the necessary work-force, and the interaction of mining concerns with the landed interests of the coalfield. All these elements had particular roles to perform and interests to protect, and in the progressive development of any coalfield the interactions became more and more complex and entangled. Sometimes the boundaries between elements were blurred, as for example in the Appalachian coalfield where the railway companies became financially closely interwoven with the actual coal companies. Not unaturally with so many different interests involved, the pattern of exploitation became a compromise rather than an economic ideal, and each company achieved its own blend of productive resources and inputs.

Settlement performed a function which was primarily geared to serving an economic purpose. This has often been misconstrued to imply that colliery settlement was a simple economic ancillary, created for convenience by the coal companies. Yet it can be shown for many coalfields, including south Wales, that sufficient momentum was provided by mining expansion to attract private capital from sources other than the colliery owners into the housing sector of the regional economy. Consequently, we have to add a further factor to the simple colliery-workforce association— that of private capital invested in housing and other property, raising problems of what was the most efficient way of housing the essential workforce in any given situation. These, and other sets of interrelationships, build up into a complex system, and need detailed consideration within any coalfield. The individual elements must certainly not be looked upon in isolation, but as part of one functioning *system*. An approach is needed which is capable of analysing the location and structure of colliery settlement as part of a wider set of interrelated features within the functional system which has been developed in any one coalfield, and this has been the overriding focus of attention in this paper.

## THE CHOICE OF STUDY AREA

In a study which sets out to develop generalizations about a phenomenon, the choice of area is of importance. Every coalfield is unique in its combination of physical and human attributes, so that a study of any one coalfield does not enable one to put forward an 'explanation' of colliery settlement which would be capable of explaining the features of settlement in any coalfield. No such claim is suggested here, but at the same time there exists in human geography the need for a set of generalizations about the many facts with which we deal. What is presented in this paper is not an explanation, but essentially a method of analysis and interpretation, demonstrated within the context of the south Wales coalfield.* It is hoped that the analytical method, based on fundamental relationships which are of universal significance within the context of colliery settlement, will contribute to geographical generalizations and prove capable of application, without major modification, to other British coalfields at least.

Carter, in his recent study of Welsh urbanism, is more concerned with establishing generalities for the internal structure of all industrial towns, and makes little direct comment on Welsh colliery settlement (Carter). Indeed he stresses the fact that only a negligible quantity of serious research has been accomplished by geographers on the enormous amount of settlement in the coalfield, a fact

---

*One major amendment only to the geological delimitation of the coalfield has been made; the coastal districts of the coalfield in south-west Wales have been excluded from this study. This zone stretches from Port Talbot through Neath, the northern suburbs of Swansea, to Llanelli. Although coalmining has been important and flourishing, it was always clearly subordinated to the manufacturing and commercial activities of the zone, especially the diverse metal industries.

regrettable in view of the important contributions made to the study of the economic history of industry in south Wales. In Carter's own words 'The actual process of settlement growth (on the south Wales coalfield) has received scant attention, whilst even the whole process by which mining camps have been transformed into functioning urban communities has scarcely been touched on' (Carter). This challenge is one which is well worthy of the geographer's endeavours!

In concluding this general introduction, it must be emphasized that the Paper, seeking as it does to establish guiding principles and methods of research, does not purport be an exhaustive documentation of colliery settlement in the entire coalfield. Those who seek a very detailed account of the growth of settlement in one or more parts of the coalfield will be disappointed perhaps, but it is the writer's hope that interest in the general field of study will be stimulated by the ideas suggested.

## CHAPTER 2

## SOME BASIC CONSIDERATIONS

*Introduction*

I T IS necessary at the outset to examine the major components of
the functional complex within the context of the entire coalfield
before going on to analyse their interaction in selected tracts.
Land, housing demand and supply, the collieries, and transport
will be discussed, but the level of analysis at this stage is deliberately
generalized; thus the land factor will be looked at in terms of broad
contrasts in its suitability and availability for housing across the
coalfield, whilst an analysis of the actual patterns of control and
exploitation will follow later.

## LAND

The land factor is one of special interest in the south Wales
coalfield, and its importance has persisted down to the present
struggle for leasehold reform. In the south Wales coalfield, where
good building land was usually at a premium, land was valuable not
only for its mineral wealth but also for its surface rights for property
development. The relief pattern did not control house construction
except in a very broad manner, but its severity gave a tremendous
impetus to the creation of a land market. Land itself could become
a considerable source of income; for the Llanharan Estate, by no
means well-located in relation to the major mining districts, revenue
from building leases formed over 30 per cent of total income in 1901
(Glamorgan C.R.O. D/DBJ/E550).

*Topography and land demand*

In the western coalfield west of, and including, the Neath valley
(after Howe), the more open relief and the slower pace of population
growth meant that there was less housing pressure on a greater
relative amount of good building land. In the eastern coalfield,
including the 'blind' valleys of the central coalfield plateau and the

'through' valleys of the eastern coalfield plateau, with a much larger population influx, and a relative scarcity of good building sites, land costs were correspondingly higher. Furthermore, costs of construction in the eastern coalfield more often than not involved much additional expense. As Richards states,

> The hilly nature of most of the south Wales coalfield (but espe-cially the plateau section) seriously handicapped many of the south Wales builders. In many cases in south Wales the cost of masonry was considerably higher than the cost for a house of identical cubic capacity erected in a more favourable area (Richards).

Thus in the coalfield plateau, where valley-sides of varying degrees of steepness were commonly used for house-building, the houses and terraces in such cases occupied niches cut into the mountain-side, making heavy excavation, infilling, embanking, and retaining work necessary (Plate I). To economize on land and construction costs, high density building was essential. Under such circumstances,

> The erection of a terraced house would cost considerably less than the building of a detached or semi-detached dwelling of identical cubic capacity. (Richards).

Terraced housing, therefore, became predominantly a feature of the eastern and central coalfield, and semi- and detached houses a feature of the western coalfield.

These two major types of house-unit, the terrace and the semi- or detached, provide the basis for much of the visible contrasts in in settlement form, but were by no means mutually exclusive in their distribution. Periodic downward movements in land prices, which occurred in phases of relative economic stagnation, enabled detached and semi-detached houses to enter into the eastern and central coalfield. Also, there existed at all times a minority of people in the latter areas who could afford the extra costs involved in this dwelling type, and this became particularly noticeable after 1900.

*The value of land for building purposes:* this can be illustrated from the leases and documents of the various large estates in the coalfield. In all cases it must be remembered that in the pioneer stages of colonization in any one district the major concern of the landowner was to ease the path of the coal master, particularly in really isolated areas. E. D. Lewis has shown that freehold farms were being sold at quite low prices in the Rhondda Fawr in the mid-nineteenth

century, before the steam coal boom really burst on the valley
(Lewis). In the case of the Llanharan estate, the mineral leases of
this period show that the coal masters were given a free hand in
choosing the initial site (Glamorgan C.R.O. D/DBJ). But when the
success of the venture was proved, locally or increasingly after 1860
in relation to the regional context, the scale of potential developments around any colliery could be fairly accurately estimated, and
concern over lease disposal became a marked feature. Thus a
common set of clauses in the mineral leases of the Llanharan,
Ewenny, Gnoll and Dunraven estates was a rent-free grant of land
for the colliery itself and associated headworks of about four to ten
acres; but the use of this land for housing purposes, which would be
to the detriment of revenue from building leases, was strictly
forbidden, as the lease for the Meiros colliery, 1885, illustrates:

> Yet so nevertheless that such appropriated lands as aforesaid
> (i.e. the rent-free grant) be exclusively used for the purposes
> aforesaid (i.e. colliery and associated works) or of the wayleave
> rights hereinafter granted but not for the erection of cottages or
> dwelling houses for workmen or others or in any other manner
> not hereby authorized (Glamorgan C.R.O. D/DBJ/J 891/909).

Later leases generally omitted the rent-free grant altogether. The
land for building purposes was normally situated on land leased
from the lessor of the minerals, but even here it is clear
from the evidence of leases that this was by no means a formality.
Land was an important item of income and expenditure, and
negotiations were sometimes protracted. At Nantymoel, the Ocean
Coal Company repeatedly warned the Llanharan Estate about the
over-meticulous housing standards and the short tenure which the
draft lease of building land on the Blaenogwr property proposed;
as in similar cases elsewhere in the coalfield, the coal company
threatened to seek land on other estates (Glamorgan C.R.O.
D/DBJ/I 389–499). During the negotiations, which lasted for
almost two years, the colliery was being opened out and its labour
force growing steadily. All mineral leases contained safety clauses
which professed to secure the rights of the lessor over the disposal
of surface leases provided that no obstructions were caused to the
mine-owner. The large estates, as will be later shown, were fully
aware of the advantages of channelling housing on to otherwise
almost valueless land, together with commercial developments

such as hotels and shops, and not unexpectedly competition developed which was to have an important impact on the final pattern of colliery settlement. Perhaps the most significant overall evidence of the value attached to land within the coalfield is the almost total prevalence of leasehold tenure, for usually only where land is very valuable and expensive does leasehold tenure become the rule. In the coalfield after about 1860, the exact date depending on the chronology of mining development in any district, land was only rarely sold freehold, except by small farmers owning their own land, and they were the exceptions. In the majority of cases the land was inevitably leased for building purposes, usually on a 99-year basis. The leasing of land gives to the land factor, and by extension to the estates, an extremely important role in the settlement complex which must be subjected to a careful appraisal.

## HOUSING DEMAND AND SUPPLY

The second major component in the colliery settlement complex is the demand for housing exerted by a growing population. Recent work by economic historians has shown that house-building in the Victorian and Edwardian eras was cyclical, and marked by fluctuations from time to time which can be traced in all regions. Thus D. Saul has discussed the possibility of recognising a general cycle for Great Britain (Saul); whilst J. H. Richards and J. P. Lewis have demonstrated the existence of long-wave cycles in house-building activity in south Wales with peaks which tended, after 1890, to be out-of-phase with events in England (Richards and Lewis). The presence of these cycles suggests the importance of factors apart from the crude demand for housing. As Saul states,

There were great variations in experiences between one area and another, and fluctuations did not, in my opinion, result from nice calculations of demand for housing, cost of building and cost of borrowing, but was essentially a speculative activity operating within a wide margin of error (Saul).

*Basic motivations in the supply of houses*

The provision of housing in the south Wales coalfield was dominated by this speculative, profit-geared activity; houses were essentially a source of income for a rentier class which developed and which provided the finance for house construction. As early as 1873, the Royal Commission on Coal found that housing in the coalfield involved the coal-owners to only a very minor degree, as this brief extract suggests:

> Para. 1563 Q. I should gather from your last answers that there is plenty of house accommodation in your district for the colliers and that the colliery owners have not occasion, as they have in Durham and Northumberland, to build houses for the men?
>
> A. (T.A. Wales, H.M.I. of Mines, South Wales District). No, it does not follow that there is plenty of accommodation; but owners as a rule have never found houses; people, as private speculators, build houses and let them, it may be to the company in some cases, but in most cases they let them directly to the colliers themselves, and in these cases the masters have nothing to do with the houses . . . as a rule cottage property pays well, 8 to 10%.
>
> (S.C. on Coal)

The construction of houses was thus dominated by considerations of profit, income, and security of investment. Many different expedients were developed by which capital was translated into bricks and mortar with the expressed purpose of providing houses for colliery workers, ranging from individual mortgages to the ultimate in

---

PLATE I. DETAIL OF HOUSE CONSTRUCTION AT ABERCREGAN, AFAN VALLEY. This is an example of expensive house construction under extremely severe topographic conditions, the actual terrace of houses occupying a narrow ledge cut into the steep mountain-side behind, whilst extensive masonry work is needed to form a retaining wall at the front. Mountain torrents were culverted beneath the terraces. The cost per unit dwelling was obviously far higher than under normal building conditions.

PLATE II. TERRACES ADAPTED TO SEVERE TOPOGRAPHY, ABERCREGAN. A rising 'staircase' of terraces was the usual response to severe topography, and these were built after 1910, mainly as a result of the active sponsorship of the landowning estate. For the Afan valley this was the high water mark of colliery colonization, the moorland and unadopted ash tracks being witness to the fact. Most of the houses in the view were recently abandoned in 1968, when the photograph was taken.

capitalist organisation, the joint-stock housing company which built on a large scale for rent, either directly or indirectly. The very essence of the guiding spirit of settlement construction in the coal-field is contained in the following advertisement, issued by one of the hundreds of house-providing agencies. To quote from a contemporary prospectus:

### Brithdir Dwellings Supply Company Ltd.

Capital proposed £10,000 in 2,000 shares Abridged prospectus: the company has been formed for the purpose of erecting 100 houses upon a field (the freehold of Mr. C. H. James) situated close to the George Inn railway station, on the Rhymney line. The company holds the land under an agreement for a 99-year lease at a modest rent of $\frac{1}{2}d$. per sq. yd. per annum, amounting to £68/8/4 per annum.

The houses as soon as built will be leased to the Powell Duffryn Steam Coal Company at an annual rent of £10/5/0 for each house, they taking a lease for 14 years. The roads have already been formed and the Directors contemplate commencing building operations at an early date. The Directors confidently look to this as a sound investment and hope to be able to pay a dividend of 6% (*Merthyr Express*, 1890).

Before going on to examine the broader relationships between housing demand and house construction it must be borne in mind that demand and supply were activated by the relative availability of money, on the part of proposed developers and renters.

The most important factors governing building activity are those which convert potential into effective demand; thus the demographic and investment factors governing building activity are less important than employment and income factors (Richards). Only by keeping this firmly to the fore can the pattern of house-building be understood, since its implications were extremely significant. Thus, to take one very important case, the much higher real wages which prevailed in the coalfield after 1899, and which persisted almost without break until 1921, constitutes one of the key factors of the coalfield's industrial history (Thomas). The effects of the transmission of these higher real incomes into the housing market stimulated a widespread and massive upsurge in house-building activity, accompanied by a very much higher

standard of house size, style, layout and construction. Together
these form a distinctive contribution to colliery settlement in the
coalfield.

## The broad pattern of house construction

Fig. 1 illustrates house-plan approvals for Sanitary (later Urban)
Districts chosen from different sections of the coalfield which had
varying rates of growth after 1850. The chief feature of all but a few
of the graphs is the existence of marked upsurges in activity,
separated by periods of relative quiescence. For the coalfield,
Richards identified long-wave cyclic peaks in 1875, 1896, and 1909.
On the graphs a peak in the early 1890s was present in all districts,
and particularly in the steam-coal districts, but by 1896 there was a
distinct trough in the latter. The long-wave trough of 1900, coin-
ciding with the Boer War, was the best-developed of all general
trends. In a similar manner the 1909 long-wave peak was also a
feature of all sections of the coalfield, although the exact timing
varied between 1908 in Rhondda U.D. and 1910 in Glyncorrwg
U.D.

These long-wave cycles were clearly dominant in the house-
building activity of the coalfield, in all sections, and represent
adjustment to the overall economic situation. Local cyclical trends
were also in evidence, as for example the peak of 1903–4 in Gelligaer
and Bedwellty districts, and the tremendous upsurge in activity at
Gelligaer in 1913, which ran counter to the downward trend in
most districts. The complete stagnation of activity in the northern
iron towns between 1877 and the 1890s is well-marked, and was
closely related to the outflow of population associated with the
depressed state of the iron and steel industry. Indeed, only the
districts of the steam-coal area had a really large amount of building
before 1899—elsewhere construction was slow and hesitant. This
makes the upsurge of activity after 1899 particularly important
outside the steam-coal districts, since it took the form either of a
rejuvenation of settlement, or the growth of settlement in new
territory. This upsurge formed the third and final phase in the
evolution of colliery settlement.

It is evident that settlement growth, at least after 1870, was not a
straightforward process in which only 'internal' factors such as

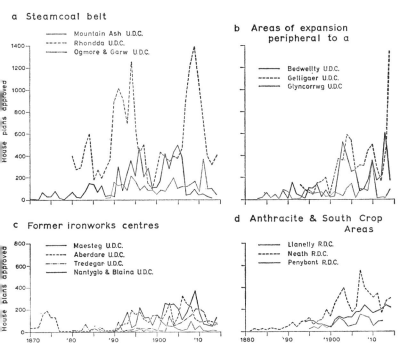

FIG. 1. GRAPHS OF ANNUAL HOUSE PLAN APPROVALS FOR SELECTED DISTRICTS. The graphs are based on tabular information presented by J. H. RICHARDS, *Fluctuations in House-building in the South Wales coalfield.* (Univ. of Wales, unpub. M.A. Thesis, 1956).

colliery sinkings or employment developments matter. Development was essentially phased—with short periods of intense and rapid house-building activity responsible for the major part. On the graphs, continuity in activity is only in evidence in the semi-rural mining zones of the south crop and anthracite field. The general state of the coal trade, and the anticipation of profits on the part of the housing entrepreneurs were the critical factors governing house construction because of the almost total dominance of the 'private' sector. This had other effects which were very significant in terms of their spatial implications, and emphasizes the need for the geographer to place his work in a relevant framework.

A less desirable but highly characteristic effect was the prevalence of overcrowding, which reached atrocious levels in newly-opening districts where the expedient of erecting wooden hutments of immense squalor was common practice. It can be shown that the total amount of house construction always lagged behind the growth of family units, and even further behind the growth of the population in total, leading to the social phenomenon of the male lodger element in the population in already overcrowded houses. After 1900 the position rapidly deteriorated to one of crisis. The economist H. S. Jevons described the situation thus:

> Housing conditions in the south Wales coalfield are not at all satisfactory as building enterprise has been quite inadequate to provide for the great inrush of population during recent years (Jevons).

The Commission of Enquiry into Industrial Unrest in 1917 estimated that a deficiency of 40–50,000 dwellings existed by 1914, and was a potent contributor to the high degree of social and economic strife present.

The provision of housing, and thus settlement, in the coalfield was therefore hampered by restraints which finally threatened the efficiency of the entire system. Essentially a 'private' sector activity, it lagged behind, in broad terms, the housing needs of the districts. In all areas after 1908, there was a marked falling away in activity by private enterprise which coincided with the hectic climax of economic development in the coal industry. The settlement component of the complex was thus becoming inefficient and inadequate to meet the needs of the coal industry which, as will be examined later, resulted in many significant developments, including the re-entry on a large scale of colliery companies into the housing sector for the first time since the 1860s. Furthermore, the friction caused by the increasing inefficiency of the housing component was accentuated by the vital spatial context to the demand-supply equation, which we shall be in a position to examine.

## COLLIERIES

The third major component of the colliery settlement complex is the colliery itself. We cannot neglect the possibility that the technology of the coal industry itself was a principal factor underlying the development of the settlement pattern, and it is important to have a foundation for our evaluation.

Most studies of colliery settlement have assumed that collieries*
present few variations in typology which might be significant for the
for the settlement response. We are concerned with collieries as
economic entities, possessing location and a capacity for employing
labour. There are only two basic types of colliery—the coal-pit and
adit. The adit mine possesses a variety of regional descriptions
within the coalfield, the most widespread being 'level' and 'slant',
but in the eastern valleys of Monmouthshire 'slope' is used; along
the south crop, 'slip'; and along the north crop, 'drift'. In this
Paper the term 'slant' is used to describe the generic type.

As a general rule, pits were larger and more expensive than slants
to develop, and this basic distinction is important for the geographer
concerned with settlement. Furthermore, a colliery has three
significant attributes in this respect—location, size (measured in
employment provided), and a certain type of life-cycle in its
development. It is further to be expected that the characteristics of
any type of colliery will alter, both in time and in area, and this
must be borne in mind.

*The coal pit*

The pit is the best-known type of colliery, probably because of its
more impressive landscape impression, a reflection of its large size.
Fig. 2 shows that the modal value for pits in the coalfield in 1913
was 600–800 men. The pit was also a relatively static entity in
locational terms, since the shafts and headgear were permanent
fixtures and the working of the seams was organised from the pit
bottom. Although pits were expensive to sink, J. H. Morris and
L. J. Williams have demonstrated that it was within the financial
means of a wide group of regional and local business and professional people as early as the 1850s and 1860s (Morris and Williams).
Yet in chronological terms, the pit usually followed the first
working of coal in any area by means of slants. Throughout the
nineteenth century pits developed in size and complexity, particularly when mining penetrated into the coalfield plateau in search
of the rich but comparatively deep seams of steam-coal. Many of the

---

*As used in this paper, the term 'colliery' denotes the area of surface ground which
is actively involved in the exploitation of the underlying coal seams, together
with the buildings and employment this represents.

earlier pits, as along the north crop, had been shallow, primitive and ramshackle affairs with rudimentary winding gear (sometimes of water-balance type) and ventilation. Because of the large capital sums involved, pits after 1850 presented considerable stability in terms of employment. At the same time, with the greater depths involved as the exploitation of the coalfield proceeded, the period of sinking and development grew longer, and usually extended over two or three years. This lengthy period usually meant the establishment of a squalid encampment of wooden huts for the sinkers at the pithead, and these were often taken over by the incoming colliery labour force for want of alternative accommodation. The reasons behind the greater stability of pits were straightforward:

It was often better to work a colliery—even at an unremunerative return on capital or at an actual loss—than stop production altogether. The shaft and works of a colliery represented capital that rapidly depreciated if all work was discontinued; without constant pumping and the regular repair of headworks and roadways a colliery could become rapidly unworkable and have to be abandoned, or restored to working order at considerable cost (Morris and Williams).

The location of pits (other than tiny pits working shallow housecoal seams or shallow seams along the north crop) was governed broadly by relief, a valley-ward location being essential in order to minimize the sinking costs. But the detailed location of the shafts, after taking account of this restraint, depended on a host of factors vital to the efficient working of the taking. As a general rule shafts were sunk to reach the workable seams at their deepest points, so that underground haulage, drainage and so on could be facilitated and cheapened. But,

. . . it is surface conditions which most often determine the precise location of a new shaft. Easy access to railways at favourable gradients, the existence of ample space for rubbish tips, buildings, screens, sidings, facilities for obtaining and removing water—these are some of the points which the engineer must consider with great care (Jevons).

Within South Wales the railway was the vital major factor: few collieries were located precisely on the valley bottoms, but were at the same elevation as the railway. The shape and size of the takings was also of great importance—it is noticeable how the congestion of collieries in the Cynon and Rhondda valleys was not repeated in areas developed after 1900. Settlements such as Merthyr Vale, Ynysybwl, Llanbradach or Bedlinog are simple cases, with one very large pit and a compact settlement grouping; this simplicity is the classic colliery—settlement model recognised by most geographers. In such cases as Mountain Ash, the larger number of collieries has resulted in a swollen size of settlement, but the intimate juxtaposition with the collieries is preserved.

The size of pits in employment capacity has caused most misunderstanding. We can see in Fig. 2 that the modal size of 600–800 in 1913 was not vast, and twenty years earlier this figure would have been considerably lower. The median size in 1913 was 1,000 men, and only ten collieries in the coalfield employed over 2,400 men. The really enormous pits described by Sorre as characteristic of the Nord coalfield were not typical of south Wales. In the latter the smaller pit was typical, and as a result the type of monolithic colliery settlement, in which the collieries were in the 1,800–2,400 class, did not occur in numbers sufficient to dominate the coalfield.

### The slant mine

The slant mine has been a neglected feature, possibly because it is often inconspicuous and ephemeral. Yet slants varied considerably in type, and have produced many problems for the geographer. The size of employment afforded by slants was small compared with pits, as Fig. 2 illustrates, the modal size group for slants in the coalfield plateau in 1913 being 1–50 men. Slants were much larger in those areas of the coalfield where slant-mining was the predominant type of mining. Thus in the Carmarthenshire anthracite field, and on the south crop, the modal range was 200–300 men. But the largest slant was still smaller than the average pit, and rarely were over 500 men employed.

Where geological conditions permitted mining by level (or slant) it was the natural method to adopt, as it needed only a small outlay of capital, avoiding the cost of sinking a shaft, and it also made for low working costs. The working costs were low because the level sidestepped the two basic problems of mining—drainage and ventilation. As the workings became too large for these

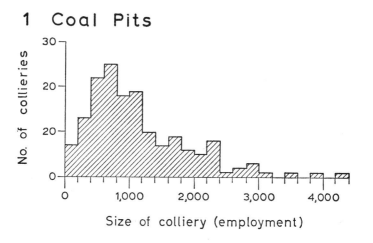

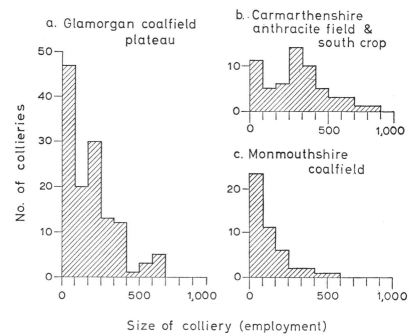

FIG. 2. SIZE CHARACTERISTICS OF TYPES OF COLLIERIES IN 1913. Employment figures include underground and surface workers, and are taken from H.M. INSPECTOR OF MINES, *List of Mines in the United Kingdom of Great Britain and Ireland,* 1913 (H.M.S.O. 1914).

problems to be solved easily, it was often simpler and cheaper to abandon them and open out a new level (Morris and Williams). This summary of the technical aspects of a slant mine stresses the key advantages of cheapness and flexibility and these had important corollaries. Firstly, the 'life' of a slant was shorter than that of a pit, and they were, by comparison, ephemeral. Secondly, it meant that the slant was confined to outcrop zones, either the major regional outcrops of the coalfield, the north and south crops; or internal outcrops of coal seams within the coalfield plateau. Thirdly, the cheapness of working and simplicity of operation meant that slants could be closed and re-opened more easily than pits. These technical factors have meant that the settlement response to slant mining has shown significant variations from 'orthodox' pit settlements.

The location of slants along the outcrop of seams meant that the entire north crop was pockmarked with the levels of the former iron companies, served by a maze of tramways. Elsewhere in the coalfield it is generally true that the slants preceded the pit in the sequence of mining colonization, the slant owners often subsequently sinking pits as their finances improved. In this way the slant and pit occurred in juxtaposition across the coalfield plateau, with the slant being in a subordinate position regarding employment. Although the ultimate employment balance in the coalfield plateau was heavily in favour of pits, the earliest colliery settlement was normally associated with the first beginnings of slant mining, an aspect of some importance in settlement location.

Within the coalfield plateau the slants mined upper, usually house-coal seams outcropping high on the valley sides, the coal being transported down to the railway by inclined tramroads or by means of wooden chutes. Before the penetration of proper railways into an area of slant mines, there were often extensive tramway systems running across the plateau interfluves to rail-heads, as for example in the area between the Afan and Neath valleys. The Neath valley marked a transitional stage in a technical sense, since moutain-side levels were systematically focussed via tramway systems at washeries or tipples located in the floor of the valley, this representing a partial concentration of activity and organization.

The slant mines in the areas outside the coalfield plateau were in lower and flatter topography; because of intrinsic geological difficulties, they were the predominant method of mining (Hare). Given rail access, the slants had a wide locational freedom along the southern outcrop, and in the Carmarthenshire anthracite field, and were generally completely intermingled with settlement, which developed a very linear, straggling form with few compact nuclei.

In these districts the slants were larger than in the plateau, and the labour demands were very high in relation to output, as a result of the extremely difficult mining conditions. Yet the sparseness of settlement, especially in the south crop zone, demonstrates the low employment capacity of this form of mining, which combined with the scattered, unstable distribution of slants, favoured a dispersal of settlement effort. The ease of openings and closures of slants, whilst economically advantageous, was not conducive to the growth of large and compact settlements: a life cycle of about six years was by no means uncommon along the south crop, although a greater stability was apparent in the anthracite field. In such areas the settlement response was cautious, because of the potentially short and uncertain nature of employment and mine. Periodic booms in demand were often met by influxes of seasonal and casual labour in the years before the Great War.

## TRANSPORTATION

The entire economic development of a coalfield is dependent on the establishment of an adequate transportation network—without this, the movement and disposal of the bulky mineral products is impossible. Within south Wales, despite the considerable if localized importance of canal and steam-operated narrow gauge tramroad, transportation was synonymous with the development of one of the densest rail networks in the British Isles. In effect this meant that railways or mineral lines penetrated into virtually every major and minor valley of the Coalfield plateau, breached high summits with incredible graded ascents, and tunnelled through inter-valley ridges. Mineral lines were constructed to lower safety and running standards than railways proper, and were consequently not allowed to run general freight or passenger services; but as far as the coal industry was concerned there was little difference. The exuberance of transportation development created problems of duplication in less buoyant periods after 1926, but there is little doubt that during the late nineteenth and early twentieth centuries the various south

Wales coalfield railway companies performed prodigious feats of mineral handling along their densely trafficked and heavily graded networks.

From the point of view of settlement geography, this dense and penetrative rail network offered opportunity not only for the movement of coal, but of men; so that besides being the vital link in the distribution of the coal production the rail network potentially had a large role to play in the adjustment of labour demand and supply, and hence settlement geography of the coal field (Jones, P. N.). Labour was a vital input in the production of coal, and it was a commodity which was more often than not in scarce supply, and rarely equitably distributed in relation to the current demand pattern for new colliery workmen. In this situation the use of the rail network played a very important formative role in the colliery settlement complex, and serves to remind us of the significance of considering this *complex* as a whole, rather than as a set of unrelated parts.

*Factors underlying the development of colliers' work-journeys*

The features associated with colliery workmen's work-journeys have had far-reaching effects on the colliery settlement pattern.*
A movement of miners from one district to another suggests an imbalance in employment and population, which was reflected in the existence of areas of predominant outflow and inflow in the coalfield.

The factors which led to the development of work-journey movements were, within any section of the coalfield, generally unique both in time and degree, but several predominant types of factors can be identified. The basic was that

. . . while at any given time the location of employment appears to be fixed, over a few years it must be regarded as fluid, and cannot be predicted with any certainty (Marquand).

---

*The term 'workmen's journeys' is used in the paper to describe diurnal movements to work of coalminers over a considerable distance; these movements were undertaken by rail, or simply by foot. There is no consistent evidence which would enable us to arrive at an accurate assessment of the latter type, although fieldwork has disclosed its presence as a relevant factor in such widely separated sections of the coalfield as the south crop, the Afan, Neath, Llynfi and Garw valleys. Unless otherwise stated, all movements referred to were undertaken by rail.

Changes in the rate of growth of colliery employment in any district meant that labour had to be attracted on a daily basis from other areas, or from districts sufficiently removed to stimulate an actual permanent migration of population.* Every colliery achieved its own delicate balance between the two, which was susceptible to rapid change. New collieries affected not only labour *en masse* but created openings for skilled men, such as firemen or overmen. This element was frequently attracted by better pay and prospects, and over a period of time developed into a complex pattern of cross-travelling, which did not necessarily reflect large imbalances of population and employment.

A second factor was the increasingly powerful attraction, for social and economic reasons, of large established settlements, which tended to slow down the rate of response of settlement to changes in employment. This factor was greatly strengthened by the mainly 'private' nature of house construction, which made the abandonment of any settlement by colliery worker or property investor virtually unthinkable.

Workmen's journeys were also fostered within the context of the south Wales coalfield by virtue of its 'dual' economic growth

---

*An intermediate type of movement in which labour was lodged free during the week, and provided with free transport home at the weekend, has also been used, e.g. Ebbw Vale Co. drew hundreds of men from the Eastern Valley on this basis in 1890 (*Merthyr Express*, 2/1890) but it was probably a transient feature in any area.

---

PLATE III. CONTEMPORARY HOUSING SCHEME ADAPTED TO SEVERE TOPOGRAPHY, LEWISTOWN, OGMORE VALLEY. This pleasing scheme of high quality houses also successfully accommodates the motor car, and provides pedestrian segregation; it has deservedly won a Civic Trust award. The photograph does not show that the estate is heated by a District Heating scheme; yet neither can it comment on the wider planning implications of continuing to invest in isolated, declining communities.

PLATE IV. EARLY COLLIERY HOUSING, TYNEWYDD, OGMORE VALE. These dwellings were built about 1870 by the Llynfi and Tondu Coal Company. Forming part of very long rows, the dwellings are small, with narrow frontages and only two front windows, and built of rough-hewn Pennant Grit. The influence of the rows of the earlier iron towns can be observed in this design and the Llynfi and Tondu Company did in fact operate important iron works at Maesteg and Tondu. All inside amenities were lacking when first built.

stages, which left a residual of large and well-developed settlements on the north crop; these were the most important nuclei of settlement in the coalfield when mining colonization was beginning in the 1850s, and they clung tenaciously to their early advantages. Even in 1935 the South Wales Industrial Survey was able to report:

The communities on the (northern) outcrop are older, correspondingly deeper-rooted, than many parts of the effective coalfield today. They have their distinctive traditions (especially a strong Welsh heritage) and their inhabitants are bound to them by close association with place and people (Marquand).

The social pressures on workers to remain in these townships as mining developments shifted south was encouraged by the iron and coal companies.

Finally, the necessary conditions encouraging mobility were ideal. The topography of the coalfield plateau channelled both settlement and railways into the valleys, so that all settlement was adjacent to at least one of the dense mesh of railways. Stations, halts and platforms were thickly distributed in relation to settlement, and many major collieries possessed their own halts.

*Difficulties of reconstructing the work-journey pattern*

In quantitative terms the movement was very considerable:
A large number of men in these valleys are already accustomed to travelling long distances to work. Thousands of men travel daily to work by train distances of from three to fifteen miles or more (Ministry of Health).

The South Wales Industrial Survey, 1935, suggested that the peak mobility of labour in the coalfield had been reached in the expansive phase before 1926, when rail transport was the only significant medium, and not later in the era of the motor-bus. Unfortunately the numerical data which is available for the pre-1926 period can only be used to give broad indications of these movements—such as the 'gradients' which existed—and does not allow us to make very refined statements.

Documentary sources have been drawn on extensively, and if the resulting information is a little disjointed, it can be combined with field knowledge to give a reasonable approximation of what occurred. The end-product of our investigations is firmly before us

in the landscape, in the form of disparities of settlement distribution
and type which are otherwise inexplicable.

Fig. 3 presents all the work-journey rail services for which
evidence has been unearthed from a variety of documentary sources,
which are indicated in the maps. The pattern is not complete, since
it is likely that many services were unrecorded in the pre-1890
period, and many railways kept very scanty details of such services,
especially of the 'contract' type. The date of first traced reference is
recorded and the services mapped according to *date of establishment;*
thus most of the services on Fig. 3A operated after 1900.

*The types of rail services provided for colliers and their distribution*

Workmen's services were of two main types, each of which had
distinctive features in relation to settlement (Jones, P. N.). 'Statutory' services were provided by railway companies for the conveyance of *bona fide* workmen at reduced fares under the provisions
of the Cheap Trains Act, 1883. The Railway Companies were not
under any legal obligation to provide such services, and only did so
when the demand was a profitable one. Statutory services were in
operation on most routes in the plateau, run by young companies
such as the Rhondda and Swansea Bay (Afan and upper Rhondda),
and the Port Talbot Railway (from Port Talbot to the Pelena,
Llynfi and Garw valleys); and also by the larger and older companies such as the G.W.R. and the Taff Vale Railway (Fig. 3 A, B).
Only the Rhymney and Sirhowy valleys were not represented of the
major valleys, and here the companies appeared singularly unin-
terested in the scope of these services.

The second type of workmen's service was the 'contract' service,
whereby the railway companies operated trains exclusively on behalf
of specific colliery companies, and for the carriage of the latters'
workmen only. These services were generally free to the colliers, and
so represented subsidized transport.

In the central steamcoal belt, as Fig. 3A shows, contract services
were characteristic of the initial phases of development of a valley,
and statutory services were characteristic of a later, more mature
phase. The sequence is seen in many valleys such as the Cynon,
Taff or Rhondda. The daily transfer of labour from established
settlements to new or remote collieries was a common feature in the

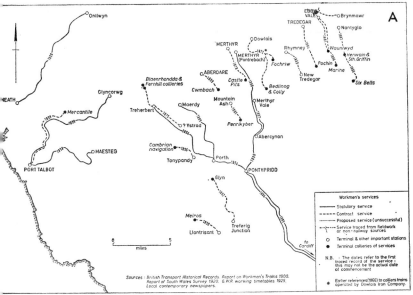

FIG. 3A. COLLIERY WORKMEN'S RAIL SERVICES IN THE COALFIELD UP TO 1899.
Only terminal points of services are shown for reasons of clarity.

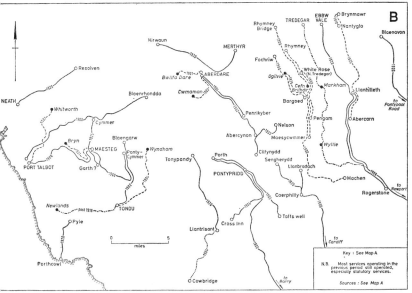

FIG. 3B. ADDITIONAL COLLIERY WORKMEN'S RAIL SERVICES IN THE COALFIELD 1900
TO 1926. Only terminal points of services are shown for reasons of clarity.

nineteenth century, with many examples being shown on Fig. 3. This type of service persisted after 1900 in the less-developed parts of the coalfield, as seen in the case of Pelena valley collieries such as Mercantile and Whitworth. Also, after 1900 especially, many companies put on contract services in order to remedy local deficiencies in the local labour supply. The services from Tondu to Wyndham, or from Aberdare to the Aman and Dare valleys are examples. In both 'initial' and 'supplementary' types of contract service the expedient of drawing labour from relatively near and established settlements with subsidized transport was both quicker and cheaper than losing production, or embarking on an ambitious housing project.

Outside the steamcoal belt there was a different relationship between service type and chronology. The valleys east of the Taff were clearly dominated by contract services. Statutory services came late, and then only to the Ebbw Fawr and Eastern Valley. Undoubtedly the causal factor in this case was the power of very large companies such as Tredegar Iron and Coal, Ebbw Vale, Powell Duffryn, to name only a few, whose financial resources enabled them to negotiate contract services on a scale which rendered statutory services unnecessary.

### *The impact of work-journeys on the settlement pattern*

In the major valleys of the central coalfield, together with the Taff, the statutory services which were dominant after 1898 were integrated with settlement in a very flexible manner. Collieries drew upon wide areas up and down valley for their labour, particularly the later collieries such as Cwm Cynon or Cilfynydd. In the upper Cynon valley and Rhondda valleys electric tramways provided further mobility, which in turn stimulated the railways to improve their efficiency—mainly *via* the introduction of steam railcars with their superior acceleration. Because of this high degree of mobility, factors such as availability of good land, general accessibility, and amenity were of major importance as settlement site factors after 1900; pithead orientation played a reduced role. A similar flexibility of settlement was present in the most intensively developed valley in the eastern coalfield, the Ebbw Fawr. Here, despite a strong flow southwards from Breconshire, statutory services distributed labour freely in both directions, from Ebbw Vale in the north to Basseleg in the south.

In many valleys of the coalfield work-journeys were essential to provide a balance between the main and tributary valleys, since the latter were often the locations of the larger, newer sinkings. The Rhondda Fawr and Cynon valleys have many examples of this type of relationship (e.g., Clydach Vale, Aman valley).

Similarly, examples of services which had the effect of greatly reducing the totality of colliery settlement at isolated sinkings after 1900 are quite common. The statutory services to the Pelena valley, Garw, Ogmore, Afan, Corrwg, Dulais, Neath and Ely valleys are of this type, and the settlement structure of these valleys simply cannot be understood unless it is realised that hundreds, even thousands, of miners travelled in from outside daily, so that the actual size of settlement bears no relationship to the employment generated. Thus Glyncorrwg U.D., 1921 Census, *Workplace Volume* recorded a daily in-movement alone of 2,230 men; this consisted entirely of 'genuine' long-distance movement. One need only to compare the small size of units such as Thomastown with the peak employment figure of over 2,500 men at the Coed Ely colliery to realise the significance of the disparity for the settlement geography of the Ely valley. Other 'drastic' examples were, of course, found in the eastern coalfield zone of through valleys; in overall terms it must be stressed therefore, that in large sections of the coalfield the relationships between the two fundamental distributions of employment and population was greatly modified, and in some cases completely distorted. The efficiency of the Tredegar Company's services was such that no colliery settlement of any consequence was built in the middle and upper Sirhowy valleys until the construction of Markham, Oakdale and Wyllie after the Great War. In the Ebbw Fach, Lancaster's Ltd. began running contract services down-valley to new and deep sinkings at Six Bells, South Griffin and Henwain, and settlement was slow to follow. The chief exception to this trend in the eastern coalfield is the Rhymney valley, where the control of the valley-head company was spatially imperfect. Thus, although the mid-valley Powell Duffryn Company put on a large number of contract services in the Rhymney valley system, it had no particular interest in preserving Rhymney-Pontlottyn, and encouraged the growth of down-valley settlement such as New Tredegar and Bargoed. In the Afan and Pelena valleys of the central coalfield, work-journeys also preserved distinctive old industrial communities in the lower valley, such as Cwmavon, and in both valleys settlement was sparse in relation to the employment offered. Statutory services later extended the labour-shed of the collieries to the coast at Aberavon and Margam.

Other examples of workmen's services utilizing reservoirs of labour exist, which are of a more specialised nature. The older mining settlements of the Blackwood Basin, where mining had stagnated after 1870, were drawn upon by contract services before the deep mines penetrated into the Basin after 1905, so that settlements such as FleurdeLis, Blackwood, Gilfach, Argoed and Pengam have a complex growth pattern. Llantrisant and Caerphilly, as two small market centres, were also utilized as mining developed around them. Undoubtedly the most remarkable attempts were those made after 1900 to draw colliery labour into the valleys from off the coalfield entirely. The most important source was Cardiff, from which services ran to the Rhondda, Taff, Rhymney, and Aber valleys. By 1902 the new Barry Railway had begun services from Barry to the lower Rhondda; this route also served a portion of the rural Vale en route to the coalfield. The proposed service from Cowbridge, the 'capital' of the Vale, had as its main aim the systemization and encouragement of an existing labour movement to the collieries of the south crop and the Ely valley.

The effects of widespread and persistent work-journey movements on the social geography and settlement structure of the coalfield before 1926 can hardly be underestimated. It is a phenomenon which has been ignored by those who saw only the insularity, isolation, and self-containment of Welsh mining communities. There was never a *fixed* pattern of colliery settlement in the major part of the coalfield, if by this we mean a relatively ordered pattern of stable communities, each having a symbiotic relationship with the local colliery or collieries. The coalfield was in a state of continual movement. Workmen's journeys by rail were in fact so deeply engrained in the economic and social life of the coalfield by 1920, that the Report and recommendations of a major government survey of housing needs in the South Wales coalfield were based completely on this factor (Ministry of Health).

We have now examined some of the most important aspects concerning the major elements of the colliery settlement complex in the south Wales coalfield. In the following chapter we shall proceed to develop some generalizations about the growth and form of colliery settlement in the most important area of the coalfield, the central coalfield Plateau.

CHAPTER 3

A MODEL OF COLLIERY SETTLEMENT GROWTH

THE NEED FOR A MODEL

THE vast size of the south Wales coalfield in itself precludes any attempt on the part of one person to investigate the growth of colliery settlement in all sections, in the depth of analysis considered necessary. But the problem of sheer size is less significant than the ultimate aim of the research, and in this respect it was felt that a useful purpose would be served by synthesizing the results of detailed investigations in more limited areas, and subsequently to present generalizations about the growth of colliery settlement, and the processes involved. Because of the need to make such a synthesis as basic and as general in its applicability as possible, an abstract model has been devised whose sole purpose is to act as a vehicle for the expression of the synthesis. It contains the essence of colliery settlement in the south Wales coalfield plateau, and attempts to fit all the elements of the colliery settlement complex into a logical framework, against which developments in any particular area can be understood in a more meaningful manner.

Perhaps nothing demonstrates so graphically the singular characteristics of colliery settlement in south Wales than an attempt at classification. Generic classifications are hardly feasible in a coalfield where settlement ranges in size from cottage groups or a single terrace, to the enormous contiguous mass of settlement in the Rhondda and lower Taff valleys. In fact, within the central and eastern coalfield the only unit of areal analysis is the *valley*, the significance of which has long been apparent in local, customary usage. In the central and eastern coalfield the terminology of classical settlement geography patently has very little relevance—hamlet, village, town are quite mis-placed in this context, and indeed this has long been recognized by many observers. Perhaps the core of the problem is the fact that, in the plateau section of the coalfield, colliery settlement of a scale and density comparable to the great coalfields of Western Europe has been imposed upon an environment initially almost as wild, difficult and uninviting as that of the Appalachian Mountains of the eastern United States.

*The construction of the model*

The structure of colliery settlement in the coalfield can only be understood in terms of its genesis, and consequently it was felt that an attempt to formulate an abstract model of settlement growth would serve to synthesize the many elements which were relevant. The main objective of the model is to assimilate all the main units of the colliery settlement structure into a comprehensible and uniform method of organization. It is strictly based on the valleys of the central coalfield, and although it is capable of adjustments according to size, special modifications were necessary in order to accommodate the Rhondda valleys and the larger iron-towns of the eastern coalfield.

The conceptual frame-work presented by the model enables a synthesis to be made of the growth of colliery settlement in the only basic organic unit within the plateau, which is the *valley*, usually having one or more smaller, dependent tributary valleys. The seemingly endless and amorphous agglomerations of settlement which fill so many of the valleys of the plateau, when approached in this way, reveal an orderly pattern of growth based on factors of relative size, date of development, spatial characteristics and pit-head relationships. A categorization of all the separate elements involved would be ideal, but the terminology would be laborious and the amount of qualification necessary would be intolerable. The model can, however, be broken down into four stages, which makes comparative studies easier; and it also relegates the question of settlement size into a relative role—in other words, only the scale of the differentiation between component units is important, not their absolute size, so that each valley-unit in the plateau can be 'fitted' according to its own size values. The model has been kept free from complexities such as a comprehensive sub-division of house-styles, because over-elaboration would reduce its value; more detailed divisions can be arrived at in the field in an intensive survey of any valley-unit. The broader elements of the relationship between the model and the townscape are delineated in the paper, but in outline only.

The construction of the model has been made possible because of the prior demonstration of the existence of certain relationships between the elements of the total settlement complex. Whilst we

cannot present the whole of the analysis here we shall endeavour
to present two important aspects—the basic premises developed by
the analysis, and illustrations chosen from the analysis to demon-
strate the major features of the model itself (Jones, P. N., Thesis,
1965).

### The areal framework of reference

The analysis was conducted in terms of two major items. In the
first place, there was the actual development of the settlement
complex—housing, transport, collieries and so on. Secondly, the
analysis focussed particular attention on the *processes* involved in the
construction of colliery settlement within the coalfield. The focus
of investigation was the coalfield plateau, particularly the Glamorgan portion. In this section of the coalfield the great mass of colliery
settlement was established, the complexity of settlement development was greatest, and here the dominance of coal reached its
zenith. The analysis was conducted within sample tracts, which were
*not* selected in a random manner, but were chosen after preliminary
investigations of cartographic evidence and an assessment of the
availability of key types of documentary evidence, which are
described later. Because this documentary material was not
uniformly available for the coalfield, and because any real depth of
analysis was impossible without it, the availability and occurrence
of this material imposed a rigorous restraint on the selection of areas
for detailed investigation.

*The central zone*, the central coalfield plateau, is the core area of
the coalfield, and the 'image' of the south Wales coalfield. 'Outcropping grits of the Pennant Series dominate the landscape, which
is characterised by flat-topped blocks of moorland separated by

---

PLATE V. POST-1900 SPECULATIVE HOUSING, PRICE TOWN, OGMORE VALLEY. The
South Wales coalfield rapidly became the housing preserve of the property investor
and speculative builder, who catered for rapidly-increasing prosperity after 1900.
This terrace is fairly representative of this prosperous phase—with wider house
frontage, larger windows, and the use of bay-windows, in this instance single.
Forecourts, as in the example, were common, along with brick facings to doors
and windows. At the back of the house was normally an extension containing
kitchen, larder, wash-room, W.C., and often a bedroom above. A great variety of
individual styles, building materials, and dwelling sizes, distinguish this phase.

deep river valleys' (Howe). All the valleys end 'blind' in the upland mass, their only outlets being southwards. In this originally very sparsely-populated, harsh environment, steam coal mining was almost solely responsible for subsequent settlement, with the exception of the upper Llynfi valley, where an iron industry developed on the basis of ironstones occurring in the Lower Coal Series.

*The eastern zone* is similar in physical character, with the main exception that all the valleys break through the Pennant Grit uplands and debouch in to the high-level strike depression of the north crop, or the 'heads of the valleys'. The rich blackband ores and coking coals of this outcrop zone formed the basis of a string of ironworks which were established in the late eighteenth and early nineteenth centuries, and formed the hub of towns such as Aberdare, Merthyr-Dowlais, and Ebbw Vale, to name but a few. Only after 1850 did mining activity move south into the coalfield plateau to any extent.

The coal types and mining conditions of the central and eastern coalfield were not uniform. Along the south crop the seams dipped very steeply, creating difficult mining conditions which delayed major exploitation until after 1900. The asymmetrical syncline structure meant that the valuable coal seams were found at greater depths in the southern parts of the coalfield (North). In the Blackwood Basin, a downfaulted depression preserves Upper Coal Series rocks, and the chief coal seams of the Lower Coal Series are found at depths, so that this area was the last to be fully exploited in the coalfield. Observations have also been drawn from a section of the south crop, and from a large tract which forms a transect from the north crop depression south into the Pennant Grit plateau.

## THE BASIC PREMISES OF THE MODEL

As shown in the diagram the analysis involved two fundamental stages—growth and process; and a number of basic premises were established which must be put forward in order that the model can be understood in full.

### Phases of colliery settlement growth

Three periods of development were recognized for this purpose, each being termed a 'phase'; their basis cartographically is the

## THE MODEL

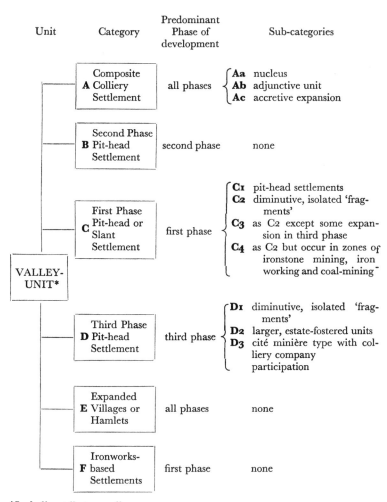

| Unit | Category | Predominant Phase of development | Sub-categories |
|---|---|---|---|
| VALLEY-UNIT* | **A** Composite Colliery Settlement | all phases | **Aa** nucleus<br>**Ab** adjunctive unit<br>**Ac** accretive expansion |
| | **B** Second Phase Pit-head Settlement | second phase | none |
| | **C** First Phase Pit-head or Slant Settlement | first phase | **C1** pit-head settlements<br>**C2** diminutive, isolated 'fragments'<br>**C3** as C2 except some expansion in third phase<br>**C4** as C2 but occur in zones of ironstone mining, iron working and coal-mining |
| | **D** Third Phase Pit-head Settlement | third phase | **D1** diminutive, isolated 'fragments'<br>**D2** larger, estate-fostered units<br>**D3** cité minière type with colliery company participation |
| | **E** Expanded Villages or Hamlets | all phases | none |
| | **F** Ironworks-based Settlements | first phase | none |

*Including tributary valleys.

complete coalfield coverage available of the Ordnance Survey 1st
Edition Six-inch Series, which was surveyed in south Wales in the
years 1876–78; the Revision of 1898–1900; and the further Revision
of 1918, with field checking for settlement built 1918–26. The dates,
though fortuitous in themselves, have great relevance in the growth
of the pattern of colliery settlement. The year 1875 saw the establishment of the Sanitary Districts in the coalfield as a whole as a
result of the Public Health Act of that year. These districts had
powers to introduce and enforce building by-laws to control
housing development, whereas prior to 1875 only certain of the
older-established settlements of the northern outcrop had any
by-law regulations, these stemming from the provisions of the 1848
Public Health Act and the 1851 Housing Act. However, even the
Sanitary Districts were not *obliged* to apply for by-law powers for a
considerable number of years after 1875, so that the actual dates of
commencement of the by-laws in the various parts of the coalfield
are not uniform. In some parts they were not introduced until the
late 1880s. The Public Health Act of 1875, which created the Model
By-laws, greatly strengthened the degree of control over house
construction, and formed the basis for 'by-law' housing as a national
type. The 'pioneering stage' in the coalfield had therefore come to a
close generally, but the colliery settlement recorded in the six-inch
1st Edition belongs to the unregulated phase of initial mining
colonization. The intermediate period from 1878 to 1898 was,
especially in the steam-coal areas of the coalfield, in the central and
eastern coalfield plateau, a phase of major economic expansion and
settlement construction, although elsewhere activity was on a more
modest scale. Finally, the third phase from 1898 forward was a
period of general high activity in house construction coincident with
the peak development of the coal industry, when new factors such as
higher wages and living standards were also becoming important
considerations. Each phase is therefore distinguished by a certain
individuality which makes them suitable for use as a basic analytical
tool. Each phase was not marked by complete uniformity—in particular the eastern section of the coalfield, with its well-developed
transport networks and the early nineteenth century growth which
accompanied the iron industry, passed through the pioneer phase
at an earlier date than elsewhere, and by the late 1870s many
features, more characteristic of the second phase in other coalfield

areas, were already present. But apart from this major tendency, no really large variations from the basic three-phase division existed.

## *Importance of the growth phases*

The purpose of the analysis of settlement development must be clarified. It was not concerned with mapping the growth of settlement for its own sake, neither was it interested in showing that some areas experienced phases of overall building stagnation or boom, since this can be shown in simpler fashion. In essence, it was undertaken to observe the *order of development of the settlements* so that the general trends which will help us to understand the settlement structure as it exists today could be isolated. A number of problems suggest themselves, such as the linkages between the pre- and post-by-law settlement; or the precise nature of the location, scale and plan of settlement units in the various phases of growth. An evaluation of the characteristics of each phase was needed so that its contribution to the whole could be appreciated. But it was also essential to make a continuous appraisal of settlement growth in relation to the development of coal-mining itself, so that the influence of changes in the location and size of colliery employment could be understood. The basic unit of analysis in terms of settlement was the individual block of houses—whether row, terrace or street. Detached or semi-detached houses have inevitably been subject to some degree of generalization, however, in the final mapping.

## *Major features of the growth phases*

*The First or 'Pioneer' Phase* of settlement growth, approximately before the mid-1870s; the dominant theme was the extremely close, symbiotic relationship between collieries and associated settlement, which found expression in the settlement pattern in two ways. Locationally, this relationship was always a very close one, although this is not to suggest that certain irregularities did not occur. This holds good whether we are referring to very small or much larger settlements, and was directly related to, and responsible for, the second observed inter-action, which involved the size of the settlement unit. Understandably, in a period of pioneering effort in a region of difficult and indifferent communications systems, the size of colliery settlement units was more or less directly proportional to the employment capacity of the colliery itself. Other features associated with the pioneer phase are the generally unsatisfactory amenity standards of much of the housing built, and degrees of remoteness which in many cases appear intolerable by our standards. *The later phases*; the dominant tendency was for the early simplicity of the size and locational relationships to break down as the whole mining complex became more developed and more flexible. But in addition, other very important factors were also operating.

In the first place, the appearance of building by-laws in most parts of the coalfield from 1878–88 onwards enforced certain standards of density, ventilation, sewerage, water supply and access which became more rigorous with the passage of time. This was an important contributory factor in the tendency for larger units of settlement to develop in these phases, which was paralleled by a relative scarcity of the smaller units characteristic of the first phase. In the second place, there was the common occurrence of large, compact and regularly laid-out additions to many existing mining settlements. These additions were often given distinctive names, and can be described as *adjunctive* extensions, as opposed to the more limited infillings and ribbon growth which occurred and which we have called *accretive* extensions. The regularity of the adjunctive extensions strongly suggests the enforcement of overall control by some party. Thirdly, and generally within the second phase only, came the creation of huge, concentrated grid-plan units at new pitheads, as at Mardy and Caerau. A progressive lag of settlement growth at pithead locations was also apparent, but particularly in the third phase, when the diminutive size of many of the pithead units suggested a parallel with the isolated rows of the first phase. Finally, the enormous influence of daily movements of colliery labour to work was an ever-present theme in all areas.

The sample studies of settlement growth demonstrated the need for extending the analysis in two directions. Clearly the actual 'process' of settlement growth needed close investigation—by process meaning the decisions which influenced the location, site and form of settlement development. Furthermore, the totality of colliery settlement was obviously unintelligible without allowing for the phenomenon of daily work-journeys of colliery labour.

*The process of colliery settlement development*

The term process was adopted because it best fitted an analytical study of the *decision-making* sequence by which settlement on the coalfield was constructed, from its inception to its completion. The analysis was concerned with the sequence of events which resulted in the physical act of house construction, and the ways in which these events have influenced the structure of settlement as it exists today. It was 'realistic' in that it sought to understand a landscape phenomenon through studying in detail the conditions which brought it into being.

No attempt was made to analyse the detailed growth of specific settlements, valuable though this would be in reconstructing the social and physical environment of the expanding settlements. Rather was it directed at the basic causes of the variations in settlement disposition and structure, 'townscape', plot arrangement and density, all of which give the ultimate 'texture' to the settlement pattern. In order to accomplish this it was essential to devise some notation by which the structure of colliery settlement could be mapped. The powerful pressure of a generally high effective demand for housing, and the relative scarcity of land, which form the vital backcloth to the growth of settlement, have already been discussed. From a study of leases, correspondence and other records of colliery companies, estates and their agents, much valuable information was acquired, but it was inevitably specific, referring to individual localities at particular periods in time. However, provided there is at least one significant factor in this set of inter-relationships which can be traced and recorded with accuracy and consistency, then the findings which emerge from these documentary sources could be applied to other parts of the coalfield. This problem was ultimately resolved by analysing for a large area of the coalfield the Registers of Deposited Plans, from which, in the opinion of the writer, the best expression of the balance between land, housing demand and employment is obtained—namely, the agencies of house provision.

*The agencies of housing provision*

The registers of deposited plans were kept by all sanitary districts from various dates, and later by the generally coterminous Urban or Rural districts. They usually record the location of any proposed development of houses, shops, chapels and so on, together with proposed layouts of new streets for estates. The date of application, and approval or refusal, the name of the applicant, and the number of houses were normally recorded; the precision of the locational information varies considerably in the earlier registers. In some cases the owner of the land was also recorded. Many authorities have even kept virtually intact all the original plans of house and estate developments, which were also of great value. From the registers it is possible to identify readily the agencies of housing provision operating in the coalfield, and in the temporal and spatial variations in the activity of the agencies was synthesized the vital balance between the components of the settlement complex.

> The erection of houses in any one district took place in order to satisfy the effective demand for houses expressed in the area, but the agent of supply differed considerably within each neighbourhood and the predominant method of supply differed from region to region (Richards).

We might also add that the importance of the agencies of housing provision also differed *within* many areas in a time sequence. Following Richards, several agencies of supply were identified: individual owner-occupiers occupying contract-built houses; investors in property and speculative builders building houses for profit, either for sale or letting—in most instances directly to the colliery workmen, but also indirectly if the colliery company itself took out leases and then sub-let; building clubs; colliery companies or proprietors; philanthropically-motivated building companies; and local authorities. Each of these agencies will be discussed, and particular attention given to attitudes which have possibly influenced the settlement pattern, so that the reader will be conversant with the implications of the various agencies when the model is described.

*The individual owner-occupier* was rather rare in the coalfield plateau during the period of study. The finance for the house came either from personal sources, or from a privately arranged loan or mortgage, and the house was built to contract by a local builder. In the registers this agency can be readily recognised because plans were always submitted for one, or two, or at the most, three houses. The townscape expression was a detached or semi-detached dwelling, or perhaps a small block of three or four houses. Considerable financial status was necessary to afford the higher land and building costs, and to obtain a private loan, which was entailed with a detached or semi-detached house in the coalfield plateau in periods of normal economic activity after 1880.

> The importance of the owner-occupier who has his own house built is much greater in the western sector than elsewhere in South Wales. While the terraced house predominates in the coalfield as a whole, this is not the case in the anthracite area, where semi-detached houses were built for owner-occupiers (Richards.)

Richards attributes this to the different social attitudes prevalent in the Welsh anthracite sector of the coalfield, compared with the rougher, mixed and generally less 'wholesome' flood of immigrants in the plateau. This was not a complete representation, because the higher total housing costs in the plateau made the terraced dwelling the norm, whilst, as we shall see, there was a high proportion of building club housing in the central coalfield plateau which was a reflection of the strong desire for home ownership. In a detailed context the distribution of this category of housing was quite distinctive, since because it was made up of very small schemes it was more or less bound to develop where there was already-existing access and public utilities; this resulted inevitably in infilling and ribbon growth. Also, these houses usually possessed both front and rear gardens, adding to the individuality of its townscape expression. The price of land was obviously a critical influence in the evaluation of the total costs of this type of development, due to the larger amounts of land needed.

*Private property investors and speculative builders:* in this category are included the investors in property who financed the erection of houses which were then let, the local or regional housing market being a favourite outlet for the small-medium investor at this time. Private investors could afford to engage builders to construct houses in advance of an anticipated demand. The other sector of this category was the speculative builder, who built mainly to increase the supply of houses, either for sale to property investors, or to colliery workers with sufficient means to buy for themselves; alternatively the builder with sufficient resources might enter the property market by letting his houses on his own account. According to Richards,

> The state of the (house) demand, and its influence on rents, would have an important bearing on the decisions of all but the most philanthropic of investors (Richards),

whilst in a similar manner the speculative builder built only in periods of *rising* effective demand. It should also be remembered that the private investor, whilst often and probably most characteristically operating on an individual basis, also had more organized outlets at his disposal. A constant feature was the joint stock property company, which financed the construction of houses for letting, and these obviously gathered their finance from the whole of south Wales. These joint stock building companies have also been included in this private speculative category.

The type of housing provided by this agency was almost invariably the terrace, especially in the plateau; after 1898 the quality of the housing varied considerably because different price levels of the housing market had to be catered for as income differences increased. It was the most widespread of the agencies in its distribution, and also the most sensitive to variations in demand. A further feature was that, because of the considerable capital reserves which property investors and speculative builders could usually draw upon, this agency could command the land market in times of severe pressure created by a prolonged period of rapid economic development. The schemes can be recognized from the Registers by the generally large scale of submissions, normally 20–40 houses, although instances often occur of very much larger schemes of 100 to 200 houses.

A mixture of owner-occupier and smaller speculative schemes was a characteristic feature in the coalfield plateau during periods of economic quiescence. This resulted in a very untidy form of development, generally ribbon growth, with often a pronounced lack of continuity in building line. In the maps accompanying the text this form of development can only be shown on a schematic basis.

*Building clubs:* These constituted a distinctive if controversial element in the settlement structure.

Fundamentally, a building club was an association of potential owner-occupiers; in addition certain 'distinguished gentlemen' of the locality were often invited to become shareholders in the club in order to give it a certain amount of prestige and stability (Richards).

Thus in the Llanharan No. 1 Building Club the local landowner, Mr. Blandy Jenkins of the Llanharan Estate, held three out of twenty shares, and acted as guarantor. The building club was normally formed either by men working at the same colliery, or meeting socially, as commonly at the same public house. According to Richards their distribution was limited to the central and eastern coalfield, but in fact there were exceptions to this generalization. The building clubs represented the tangible expression of a strong desire for home ownership within the plateau sections of the coalfield where collective action was essential if the thriftier miners were were to provide their own accommodation against a background of inflated land prices and high construction costs.

> The clubs were financed on a share basis, the value of the share being equivalent to the local unit cost of erection of a terraced house plus a small sum which would cover administrative costs (Richards).

The shares were paid-up monthly, subscriptions usually beginning prior to the commencement of building in order to provide an advance to the contractor. The trustees of the club held all the individual leases and conveyances of the members, and on the strength of this they negotiated a mortgage with building societies, west of England societies often featuring prominently in these affairs. Only when the shares had been fully paid-up, after five to fifteen years, were the building clubs wound up, and the deeds distributed. The schemes are easily recognized in the registers since they were specifically distinguished. They are far less readily distinguished in the field, but are always the most monotonous in their uniformity, and almost invariably terraced; the uniformity was in fact imposed by the rules of the building clubs. Thus rule 3 of the Abergwynfi No. 1 Building Club states:

> All houses connected with this society shall be erected according to one and the same plan and specification, and in the same style (Glamorgan C.R.O. D/D Ra 14/18).

The specifications of building club houses were also generally rather austere. Furthermore the members of building clubs were often employed at the same colliery and therefore had a strong

locational preference to be near that colliery when a site for poten-
tial development was being considered. Thus rule 23 of the
Cilfynydd Building Club, established by workmen at the Albion
Colliery, Cilfynydd, in 1886, stated:

> The Committee of Management shall, as soon as practicable,
> choose a suitable spot on Cilfynydd farm for the purposes of the
> Club, and the trustees shall enter into an agreement with the
> Lessors for the granting of leases thereof to the members of the
> Club (Glamorgan C.R.O. D/D Vau Box 4).

This strong element of locational commitment is of great significance
in the types of settlement units produced by building clubs. The
proportion of building club houses in the total dwelling stock was
considerable but never as high as the South Wales Regional Survey
suggested in 1920—

> The most usual agency for the erection of dwellings in South
> Wales during the past half-century has probably been what is
> known as the Building Club (Ministry of Health).

The general ratio in the central coalfield plateau for the post-1878
period was about one-quarter; before this date the proportion would
have been very small, except possibly in the Merthyr and Aberdare
districts.

*Colliery companies and coal owners:* in the post-1878 period the
registers indicate that their importance was strictly limited.
However, before 1878 their role was much greater, especially in the
more isolated districts, and it is apparent that the slight participation in the second and third phases was due to the efficiency of other
agencies.

> With the exception of 1914 the recorded number of houses
> approved each year for colliery companies was low (Richards).

In this way, Richards hints at the interest expressed in housing by
colliery companies in the later years of the third phase, when the
colliery settlement pattern had almost been completed. As Jevons
suggested in 1916, this re-entry into the house market was due to the
rapid decline in the ability of the traditional agencies to provide
sufficient accommodation in the peak years of the coal industry.

> The failure of the Building Clubs and other building agencies
> after 1906 to provide an adequate supply of dwellings compelled
> many colliery companies to undertake building schemes (Ministry
> of Health).

This failure was attributed to the rapid upward spiral of land costs, labour costs and house construction costs, which squeezed many of the agencies, especially the building clubs, into a state of reduced activity and so forced many coal owners to reconsider their own position. This had significant implications for the overall settlement pattern since only a minority of the companies were really affected, and these all had strong locational ties to their particular collieries. Before 1906 the participation of the coal companies was very irregular and tended to be un-systematic; many companies played a small but consistent role, but in the main the degree of activity was normally related to the isolation of the new development. Above all, it must not be assumed that a lack of participation denoted an uninformed and irresponsible attitude towards the housing function, which was by no means the truth.

Contrary to what has often been thought, in any period of rapid and unplanned industrialization a specific labour force was in fact less tied to a locality if it was living largely in company built or owned houses, because the men had no financial commitments in the area. Thus in 1894 when a stoppage affected the South Dunraven colliery in the upper Rhondda Fawr, 46 out of a total 100 company houses were vacant in under four weeks (N.L.W. Dunraven Documents, MS 489). The extra capital costs incurred were also a deterrent, but perhaps the advantages and disadvantages of entering the house market can best be illustrated with reference to the experience of one major company, the Ocean Coal Company. In 1885 the Company had entered into an agreement with a Cottage Company (a joint stock concern) for the supply of one hundred dwellings at the new sinking in the virgin Dyffryn Clydach valley, which was later to be named the Lady Windsor colliery. The houses were to be ready for occupation by the summer in order to receive the first influx of workers for the new mine, and the houses

---

PLATE VI. TYNEWYDD, OGMORE VALLEY: AN **Aa** UNIT. A general view of an early **Aa** unit built by the Company to serve high-level slants and a deep pit further up valley. Garden provision is not generous for this period. The main valley railway line is on the foreground (opened 1865), whilst the larger, more elaborate dwellings in the foreground were probably built for colliery overmen. The extensions behind the houses are prefabricated blocks installed by the Local Authority, and containing bathrooms etc.

were let direct to the Coal Company, which was thus responsible for rents. Beyond this point the Coal Company would not go, despite the fact that the Cottage Company was ready to build more houses under similar terms. A letter from the Coal Company to the site general manager gave the reasons:

A colliery company has much to gain by encouraging the public to build largely in its vicinity so as to secure their keep and support in any conflict they may be involved in with their workmen . . . in view of this it is worthwhile for the colliery company to do its best to encourage the public to build, even though by doing so it gives up a favourable opportunity for investing capital in a perfectly safe and very lucrative investment (N.L.W. Llandinam Documents, Section E, 297).

Thus at the Lady Windsor sinking the coal company had given the best land to the Cottage Company, and was at its own expense proceeding with all haste to build a new road to the main Taff valley so that further private investment would be encouraged. The circumstances of each company also played an important role, and it is clear that in a venturesome economic activity such as colliery sinking the majority of companies did not, or indeed could not afford to, place valuable capital in housing, despite its apparent profitability. It is most significant that in the case of the Lady Windsor colliery, a fairly representative second-phase sinking, the private property investor was prominent at the very outset. During this phase the coal companies could generally relinquish all housing responsibilities, and most of them did; examples of colliery company housing were few, and usually connected with special factors.

*Other agencies* played a minor role, and entered the field at a very late hour. The *philanthropically-motivated building companies* built working class houses of superior quality to be let at modest rents. Easily the most significant was the Welsh Garden Cities Company, an offshoot of the South Wales Garden Cities and Town Planning Association which drew its inspiration from the contemporary movement in England, and whose estates were modelled on low density, cottage-style layouts. This company built on 26 different sites after 1900, mainly in the Blackwood Basin, and sponsored a co-operative form of estate management in which the rent profits were used for the general upkeep and improvement of the estate for the benefit of the residents.

*The local authorities* entered the housing field mainly after 1919, though isolated examples do occur in the pre-1914 period after 1909. The years 1920-22 witnessed a boom in municipal construction across the coalfield which was coincident with the very generous but short-lived government financial assistance for working-class housing. The housing conformed to the standard type in design and layout, although in locational terms there were some interesting repercussions of the siting policy adopted.

## THE MODEL AND ITS PRESENTATION

The reader has now been introduced to the basic analytical methods which lie behind the formulation of a model of colliery settlement development in the plateau. An outline of the most important findings has also been given, and it now remains to describe very briefly the method of presentation which has been adopted for the next section of the Paper.

The model of colliery settlement will be presented in *stages*, and associated with each stage, by way of illustration, will be a discussion of some 'real' situations which have been chosen from a much wider range of material. The real examples are intended to amplify the major generalizations of the model, and also to present some of the details of development which could not otherwise be accommodated. The maps which accompany this section are complete in themselves, however, even if only one or two major points are being discussed in relation to them.

### *The pre-mining background*

For the model a valley of the 'blind' type so common in the central coalfield particularly has been devised. From Fig. 4A it can be seen that this is deeply-cut into the generally level plateau surface of between 1000 ft to 1400 ft. The main valley is joined by a number of tributary valleys of varying size, that on the west bank being essentially a very precipitous V-shaped glen running steeply inland. A thin scatter of farmsteads would occupy the main valley, with some tendency to choose sites along the break-of-slope between the gentler valley-bottom slopes and the steep valley-side slopes, which also marks the general limit of enclosed land. The farms were tenuously linked by rough cart-tracks and a down-valley 'parish

## KEY TO FIGURES 5, 7, 9, 11, 13.

| | |
|---|---|
| ++1863++ | Railway or mineral line. Dates refer to opening. |
| ⊏⊐ | Settlement in existence 1875-8 (First Phase). |
| • • • • • • • | Settlement built between 1876-8 and 1898-1900 (Second Phase). |
| x x x x x x | Settlement built between 1898-1900 and 1926 (Third Phase). |

**Pit**     **Slant**

| | | |
|---|---|---|
| ● | ▲ | Present in first phase. |
| ⊕ | △ | Opened in second phase. |
| ⊗ | △ | Opened in third phase. |
| ◯ | △ | Working throughout following phase. |
| ◯ | △ | Closed during following phase. |
| ◯ | △ | Re-opened during following phase. |

## KEY TO FIGURES 4, 6, 8, 10, 12

| | |
|---|---|
| ++++++++ | Railway or mineral line. |
| ▬▬▬ | Houses built for colliery owners (except Fig. 4E)* |
| • • • • • • • | Houses built for property investors and/or as speculative ventures. |
| o o o o o o o | Houses built for building clubs of colliery workers. |
| x x x x x x | Houses built for owner-occupiers. |
| x • x • x • x | 'Mixed' housing development of owner-occupier category and small speculative schemes. |
| ⊥⊥⊥⊥⊥⊥ | Houses built for local authority (pre-1926 only). |
| ⊏⊐ | { Pre-register settlement in Figs. 6, 8, 10, 12 <br> { Houses built in previous stage of model in Fig. 4 |

FIGURE 4 ONLY

**Pit**     **Slant**

| | | |
|---|---|---|
| ◯ | △ | Working throughout particular stage of model. |

\* In Fig. 4E (classification of settlement) all settlement is shown in solid black.

INDEX DIAGRAM TO MAP SEQUENCE OF GROWTH OF COLLIERY SETTLEMENT AND AGENCIES OF HOUSING PROVISION IN SELECTED TRACTS. This diagram serves as a composite index to Figs. 4–13 inclusive, but it will be noted that some amendments are necessary for some figures, especially Fig. 4.

road' towards the major urban settlement of the area. An inn, church, and a tiny group of shops cum dwellings would complete the settlement of this valley system in the period before mining colonization. In the model, the entry of mining is directly related

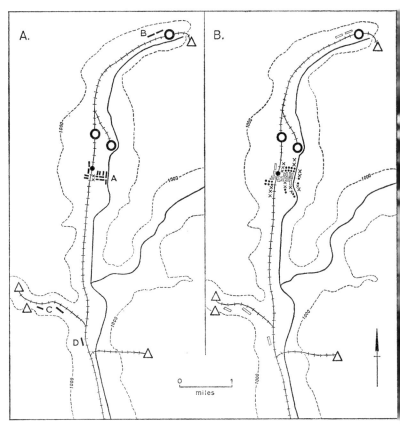

FIG. 4. A MODEL OF COLLIERY SETTLEMENT DEVELOPMENT. **A**—Stage 1 ; **B**—a sub-stage of less rapid growth. It should be noted that the pre-mining settlement pattern is not depicted but the parish roads are shown as solid lines. The letters refer to features mentioned in the text.

to the construction of the standard gauge railway, but in reality this was not always the case. As in the coalfield of north-east England, extensive tramroad systems of high capacity existed in

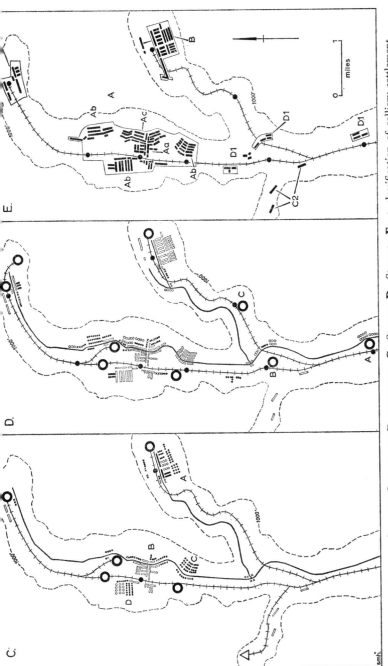

FIG. 4. A MODEL OF COLLIERY SETTLEMENT DEVELOPMENT. **C**—Stage 2; **D**—Stage 3; **E**—a classification of colliery settlement in the model. The letters refer to features mentioned in the text. Parish roads shown as solid lines.

many valleys from the early decades of the nineteenth century, and on which steam motive power was quickly introduced. This was especially the case with the eastern part of the coalfield, and it would be quite erroneous to attach any deep significance to the conversion of these 'tramroads' into statutory railways in the 1850s and 1860s. However, in the remoter central and western plateau regions some use was made of interfluvial, low capacity tramway networks, of which the most extensive exploited the area between the Neath and Afan valleys, and led to canal-side tipples on the Vale of Neath Canal. In terms of our model, it is therefore necessary to qualify the main determinants of the initial coal mining phase: in many valleys of early (i.e. pre-1850) exploitation, this was quite likely to be a high capacity *tramroad* rather than a statutory railway. In the model, no coalmining as a commercial activity would have occurred before the coming of the railway, though this would not preclude the possibility of small surface excavations for domestic uses. In addition, we shall postulate for our model the existence of a larger urban settlement about 3 to 5 miles beyond the southern limits of the map—this settlement to be dependent either on service functions or on industry, particularly an iron industry. This urban settlement has been assumed in order to provide a source of supplementary labour; in reality, an adjacent coal mining valley in a *more advanced* stage of development would serve equally well.

## THE FIRST STAGE OF THE MODEL

In the first stage of colliery colonization, at a date of development between 1850 and about 1875, a railway has been constructed along virtually the entire length of the main valley to its head, and a large number of collieries of different types have begun production. In Fig. 4A, as in others of the series, the colliery settlement has been classified by agency of construction. There is a very close relationship between the collieries, whether pit or slant, and actual settlement, and this has resulted in settlement construction in isolated locations, particularly in the tributary valley. The pits in mid-valley, with their larger employment capacity, have attracted a more substantial settlement response, and a compact settlement has grown up with its major axis aligned down-slope from the main parish road towards the railway. The establishment of a compact, nucleated settlement is aided in this section of the valley by the slope and general site characteristics, although in detail its location is not adjacent to the pit head. The reasons for this could lie in specific clauses of the Minerals Lease, in the comparative prices of land for building purposes, or infrequently in the fortuitous acquisition of a freehold property by the colliery company. It is apparent that most of the colliery settlement has been built by the colliery proprietors, which is explicable in terms of the sheer isolation of the valley, and the relatively small scale of mining activity, which had not reached 'boom' proportions. Nevertheless, private investment in building has already taken place in the largest valley settlement, in the form of houses built along the main axial street. Many of these would be occupied by pioneer tradesmen, as shops/dwellings, and a sprinkling of public houses would also be present.

*Predominant house types*

The dwellings of this first stage of activity were poor in quality. Many of the isolated rows would be of single-storey cottages built of rough stone. In the larger settlements the streets would be of terrace type, the dwellings having small rooms and usually inadequate ventilation and light (Plate IV). Sewerage, water supply and other services are non-existent everywhere. The provision of garden space was usual, however, and in fact the presence of both front and back gardens was not uncommon.

This outline of the early stage of settlement genesis leaves us with a very basic and elementary set of relationships and a settlement distribution pattern which consist of a scatter of isolated rows or terraces (B, C, D) one isolated but compact valley settlement at A, and a basic network of railways and collieries. Amongst other components which would undoubtedly have been present would be a good representation of small, plain Welsh Nonconformist chapels.

*Selected case studies*

*The Afan valley system* (Fig. 5): pre-by-law settlement was very scanty and widely dispersed. It was representative of pioneer settlement in the coalfield in that it consisted of the most elementary housing accommodation for small groups of workmen and their families, and nothing else. This housing took the form of small

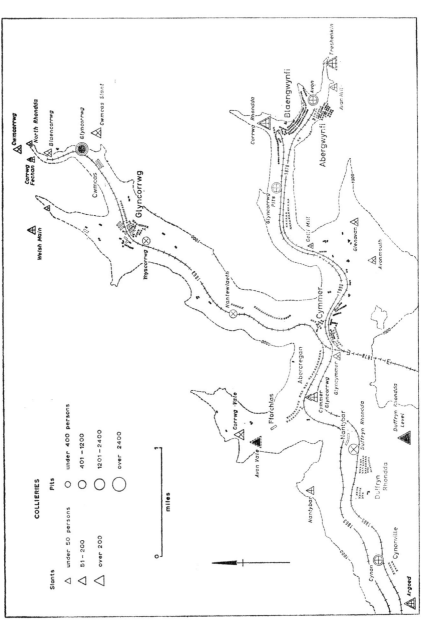

Fig. 5. Growth of Colliery Settlement in the Upper Afan and Corrwg Valleys. The size gradings

stone row cottages, generally of single-storey construction, and often in incredibly isolated conditions,* but where they were in close proximity to the collieries they served. Cwmcas, Blaencwm, Fforchlas and Nantybar are type examples, whilst Gelli cottages, although built in 1881, really belongs in spirit to this phase of colonization. These early settlement units were often badly located for subsequent growth. Thus Nantybar, built in association with the Avon Vale Colliery, was sited on a superb morphological flat, but there was little room for expansion. Fforchlas was perched on a precipitous hillside track, and Cwmcas was sited in the deep and narrow glen of the upper Corrwg. Consequently, even within the first phase, settlement relocation had taken place and colliery settlement had been constructed near the old Celtic parish church at Glyncorrwg. The amount of settlement at Glyncorrwg hamlet was still small, but its site was more favourable for expansion, and it had the advantage of serving a pit, in an area where slant mining predominated at this time.

*Garw valley:* settlement here was in a similar stage (Fig. 9). The total amount was small, and mainly related to slants. The two pits were very recently opened and still in the process of development, since the full exploitation of the valley had begun only a few years previously with the construction of the railway in 1876. Rows existed at Braichycymmer on the western valley-side, and down-valley at Llest and Pontyrhyl; all were very isolated from the nearest urban centre at Bridgend. However, the pattern of settlement was rather different in the Llynfi and Ogmore Fawr valleys.

*The Llynfi valley* was dominated economically by two large iron-works, both in production in 1878, and these had stimulated a considerable settlement response, of which Maesteg itself was the most important (Fig. 7). This township was aligned in strassendorf form along two sub-parallel parish roads, with side-streets branching off, with a considerable amount of ribbon development to north and south. Rows of cottages built by the iron company were grouped together at Nantyfyllon (J on Fig. 8), but other rows were scattered up the length of the valley, housing some at least of the workers at the coal and ironstone levels which served the ironworks.

---

*Blaencwm row (not shown) is possibly the best example, located on the plateau surface at over 1000 feet to the west of Glyncorrwg.

There was no element of pit-head location, the collieries and iron-works forming a girdle around the settlement, with only the Garth Merthyr and Maesteg Merthyr collieries actually intermingled with settlement. As indicating the larger scale of settlement in this valley, we can cite the example of the village of Garth, which was far larger than any unit hitherto encountered in other valleys.

*The Ogmore Fawr valley:* mining colonization had begun early in the pioneer phase, the railway having arrived in 1865. Consequently the colliery settlements in Fig. 9 were larger, although still very compact. Nantymoel is a good example of a pithead settlement of this first phase, and was planned and built by the Ocean Coal Company after 1866. A further settlement had been constructed by the Llynfi and Tondu Coal Company down-valley at Tynewydd (Plate VI), and this had expanded along the parish road by 1878. The scale of settlement unit was larger than in the Afan or Garw valleys but the styles of dwellings were similar. Concentration at the pithead was much in evidence, and there was only one small isolated unit, at Aber.

*Departures from the model*

Two important cases of initial stage development cannot be included in the model because of their dominating effect on the pattern of subsequent growth; both cases might in fact be looked upon as limiting cases, in which first stage development was either remarkable for its completeness, or for its comparative paucity. The model cannot be applied in its *simplest* form to the Rhondda Fawr, or to the major eastern valleys—although the outline classification of form elements is equally applicable to them.

*Rhondda:* the limiting condition of a rare *completeness* of colliery settlement in the first initial stage is illustrated in Fig. 11, depicting the growth of colliery settlement in Rhondda, although for clarity the lower sections including Porth, Hafod and Penygraig are not shown.

In 1876 one factor was quite different from other valleys in the coalfield plateau—the almost complete distribution of collieries throughout the length of the valleys. Mining colonization had reached an advanced stage, and even tributary valleys such as the Clydach had been opened up. The outstanding characteristic of the settlement pattern of the Rhondda Fawr was the ribbon of settlement along the winding parish road, which can be compared with the extremely straight valley bottom course of the railway. Specific pithead settlements comparable to Nantymoel were uncommon, Llwynypia being perhaps the best example. Elsewhere, the ribbon of settlement had swollen out where collieries were concentrated to any degree, such as at Treorchy, or Pentre. There were still considerable breaks in settlement continuity, such as between Treherbert and Treorchy, or between Ystrad and Llwynypia. A grid-plan was followed in the larger units, the gentle valley-bottom slopes being easily utilized for this purpose. Most colliery settlement in the valley in 1878 was therefore not directly related to the pitheads; it had developed in a more organic fashion in relation to the main avenue of valley communication, the parish road threading its way from Pontypridd to Blaenycwm. This pattern was fostered by the completeness of the distribution of the collieries within the valley.

Nevertheless, settlement 'fragments' and larger pithead units occurred everywhere outside the main valley. Isolated rows were located at Fernhill pits, Abergorki, Blaenrhondda, Blaenycwm, Blaenllechau, and in Cwm Clydach, which were comparable in form and function to Cwmcas or Pontyrhyl on previous maps, although as a result of the overall development of the main valley the degree of isolation was less pronounced.

Cwmparc and Ferndale were large pithead units, linked to the adjacent collieries; along with Nantymoel and other units such as Mountain Ash in the Cynon valley, they were forerunners of the large pithead groupings of later phases in that they were concentrated and compact. With the exception of Blaenllechau and Ferndale, the Rhondda Fach, as shown, was devoid of colliery settlement at this point of time.

*The main eastern valleys of the coalfield* illustrate the fact that the presence at an early stage within a valley unit of a large urban concentration (in this case based upon the earlier extensive growth of iron-production) disturbs the pattern of colliery settlement since it drastically reduces the need for additional housing if the daily movement of workmen is well organized. As a result 'degraded' examples of colliery settlement tend to dominate, although the outline classification, which considers entire valley-units, is unaffected.

Fig. 13 illustrates settlement development in a tract of the eastern coalfield, comprising the Sirhowy, Rhymney and Bargoed Rhymney valleys. In all the eastern coalfield valleys the ubiquitous element was the valley-head ironworks or group of ironworks and its associated settlement, and the valleys in the sample tract display this feature. It should be noted that the Bargoed Rhymney, along with the Taff Bargoed valley further west, fulfilled a similar function in relation to Dowlais as did the upper Taff to Merthyr, or the Sirhowy to Tredegar.

In 1878 settlement was concentrated in Rhymney, Tredegar and Dowlais at the valley heads. The economy of these towns centred on the large ironworks, with their associated collieries and ironstone mines scattered in levels and patches on the 'mountain' mineral concessions as well as in the works' precincts. Dowlais in 1878 was larger, and more compact, and like the other main companies the Dowlais Iron Company maintained an extensive private railway and tramway network connecting its collieries on the 'mountain' with the ironworks. Some isolated rows of colliers' houses were built in the wretched wilderness of moor and bog which constituted Dowlais mountain (the actual outcrop zone), as at Pantywaun, but by 1860 the Company was sending colliers' trains from Dowlais up to the collieries on the mountain, at Cwmbargoed, Fochriw, Tunnel, and Colly in the Taff Bargoed valley. In a like manner special colliers' trains were operated by Tredegar Iron Company to its more distant collieries, such as Bedwellty Pits.

Consequently the amount of 'pure' colliery settlement which had developed by 1878 in the tract was small, and owed its existence mainly to the vagaries in the geography of mine ownership. In the Bargoed Rhymney valley a few rows existed at Fochriw, and at Penybank; further down valley a small pithead unit had developed at Deri, which extended along the valley road to Bargoed Terrace in the direction of Cil Haul colliery. In addition, the Darran colliery of the Rhymney Iron Company was served by this settlement. The collieries in the Bargoed Rhymney valley were essentially coking coal collieries, and all had batteries of beehive ovens; this also applied to the cluster of collieries in the Rhymney valley—Tir Phil, White Rose, Hope, New Tredegar and Cefn Brithdir. The settlement response was varied—New Tredegar had a compact, grid-plan layout, but the most common form was the cluster of

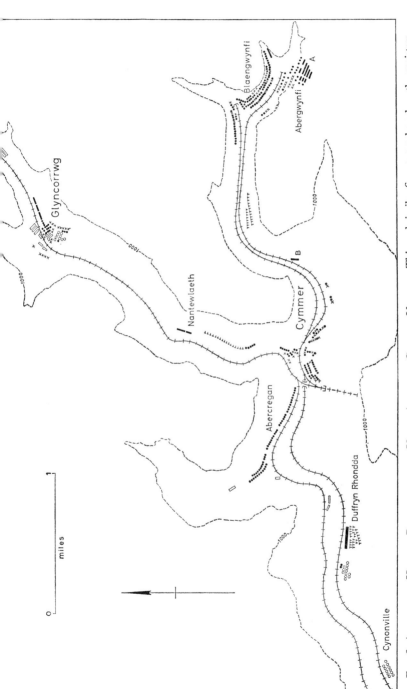

FIG. 6. AGENCIES OF HOUSING PROVISION IN THE UPPER AFAN AND CORRWG VALLEYS. This and similar figures are based on the registers of deposited plans for sanitary, later Administrative, Districts. The letters refer to features mentioned in the text. The registers date from 1882. It is possible that a small group of houses at Abercregan were built by a building club on the lines of an earlier estate plan, but the field evidence is not conclusive.

hill-side terraces, such as Tir Phil, Sebastopol and Cwmsyfiog. That this settlement developed at all was potentially a reaction to a critical gap in the Rhymney Iron Company's down-valley mineral leases, which extended into the Bargoed Rhymney, and further south in the Blackwood Basin section. The exploitation of this strategic stretch lay in the hands of a variety of independent companies, notably Thomas Powell, subsequently to be reconstituted as Powell Duffryn Steam Coal Company. In a comparable manner, Merthyr Vale–Aberfan was developing in 1878 around new 'independent' pits sunk just south of the Cyfarthfa and Hills Plymouth Iron companies' leases. In the Sirhowy the entire upper and middle, and some of the lower sections of the valley were under the control of the Tredegar Iron Company; where small breaks occurred, such as at Hollybush, a small unit had been built. The Ebbw Vale Iron Company, and the Dowlais Iron Company enjoyed comparable monopoly positions over the down valley mineral concessions.

The diminutive scale of colliery settlement in the first phase can be traced therefore to the interaction with the older iron towns, whereby these were suppliers of labour to the new collieries. The Dowlais Iron Company in 1866 explained to the Great Western Railway that '. . . with regard to the conveyance of workpeople, it must be understood that all the Dowlais Iron Company's workpeople employed on the Mountain [i.e. pits and levels south of Dowlais such as Colly, Fochriw] must be carried free . . . in fact the class of people we now carry, there must be no reservations' (Glamorgan C.R.O. D/DG Section F, Box 4, letter 9/11/1866). The final agreements to lease the Dowlais Iron Company's own railway system to the Great Western and Rhymney railways jointly, so providing direct if severely graded access to Dowlais works from Cardiff and Newport, contained therefore the following conditions:

> The Railway Companies from time to time will so long as they remain under the obligation to work the domestic traffic of the Iron Company's railways . . . convey for the Iron Company their domestic traffic and by some or all of trains employed for the purpose of that traffic and in a carriage to be provided by the Iron Company the agents and workmen of the Iron Company free of charge and at such times and in such manner as are from time to time reasonably satisfactory to the Iron Company

provided that the Railway Companies shall not be required to carry any of them on a Sunday and the Iron Company will indemnify the Railway Companies against any liability claim or demand for compensation in respect of any accident or accidents which may happen to any such agents or workmen (Glamorgan C.R.O. D/DG Section F, Box 4, Dowlais Iron Company Agreements with the Great Western Railway and the Rhymney Railway as to domestic and foreign traffic, 1872, 1877).

The Dowlais Iron company provided, or arranged to provide, free and hence subsidized transport for its colliery workmen, and there is evidence which suggests that the provision of special colliers' trains to distribute labour from the iron towns was an established feature in most of the eastern coalfield valleys, so avoiding the capital expenditure on housing provision. Thus within the tract on Fig. 13, the Tredegar Iron Company and the Dowlais Company operated such services, the former by arrangements with the L. & N.W.R.; in the Rhymney valley the down-valley collieries were not owned by the Iron company, so that subsidized services were not available. But in spite of the growth of settlement near the down valley collieries there was also a spontaneous daily movement of miners southwards by train from Pontlottyn and Rhymney to take advantage of the expanding employment opportunities (*Merthyr Express*, 1869).

The maldistribution of colliery settlement in relation to employment by 1878 was due to the development of transport services geared to the daily distribution of labour from the iron towns, which were anyway faced with declining employment in the iron industry after the 1860s. The strong social ties which had been knit, the rudimentary beginnings of urban facilities, all were potent factors increasing the cohesion of the valley-head settlements, and workmen preferred to live in these established Welsh communities than move to isolated 'pioneer' mining units.

*Land ownership and settlement form*

The landholding pattern had many relevant effects upon the detailed morphology of early colliery settlement, despite the general lack of much real pressure on land resources. The effects were most often observed in the siting of the initial nucleus within a valley-unit, and some examples will be quoted at this stage from the valleys already discussed.

Freehold land for building was especially valuable when colliery

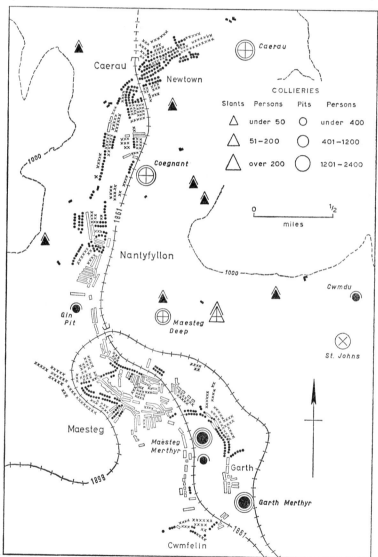

Fig. 7. Growth of Colliery Settlement in the Llynfi Valley

companies were largely having to build their own settlements. The original colliery settlement in the Ogmore Fawr valley was built after 1865 at Tynewydd (B on Fig. 10, also Plate VI), some distance south from the main pit at Wyndham, but located on the sole patch of freehold land which the Llynfi and Tondu Coal Company possessed in the valley (Glamorgan C.R.O. D/DRa 11/476–513). This served as a nucleus for later development. The price of lease-hold land for building purposes was another significant initial locational factor, and bartering over prices, or the quality of the proposed houses, was not infrequent. An early example occurred in a clash between Thomas Joseph and the Bute Estate over the proposed 'model' colliery village at Bryn Wyndham (D on Fig. 12) in the upper Rhondda Fawr (N.L.W. Dunraven Documents, Box 393).

A later example occurred in the Garw valley (Fig. 8), and again involved the Dunraven Estate. In 1877 a director of the Ffaldau Steam Colliery Company stated:

We ask you (i.e. the Estate) to grant land upon which to build workmen's cottages and a manager's house to enable us to have men live near the colliery, on the same terms as Mr. Blandy Jenkins (Llanharan Estate) and Mr. Thomas of Braichycymmer let their ground for building purposes—they charge 2*s*. 6*d*. per perch to any person requiring ground, and you are well aware that their lands are in the immediate neighbourhood of our colliery. Our wish is to get land from you (as mineral lessors) rather than go to any other landowner (N.L.W. Dunraven Documents Box 390).

The plea went unheeded, in fact, and the colliery company had to build its first houses on a tiny irregularly-shaped field belonging to Braichycymmer farm which formed an enclave into the Ffaldau property of the Dunraven Estate (K on Fig. 8); many poor-quality rows were also built on the main part of the farm on the western side of the river. Moreover, later housing was diverted to Nanty-rychen farm on the Llanharan Estate, and only after 1898 were the Ffaldau and Ty Meinwr properties of Lord Dunraven developed in earnest.

Consequently, the initial site factors of colliery settlement should be carefully considered, particularly in those valley units where a mass of later development occurred around the original nuclei.

# A SUB-STAGE OF RELATIVE STAGNATION

A sub-stage of relative stagnation could occur within any one of the *main* development stages, and indeed in all three. It represents a period of reduced activity in coal mining and house construction, and generally a period of slower economic and demographic growth. This sub-stage often occurred after a period of very rapid growth, in which the growth rate in the colliery labour forces tended to subside, so that any additional labour could normally be supplied by the natural increase of the local population (Jones, 1965). For the purpose of our model we have shown a sub-stage *between* two main phases of growth (Fig. 4B).

## *Dominance of accretive growth*

In terms of settlement form such sub-stages were marked by accretive growth around the main settlement nucleus. This took the form of 'mixed' speculative and contract-built houses, with a pronounced tendency towards ribbon development along the parish road. Small terraced streets of private speculative housing adjacent to the existing streets are also seen. Perhaps the most common feature of the house-types was the prevalence of semi- and detached houses, indicating that land prices had again become 'easier'. In valleys with more than one main nucleus the additional accretive growth would generally gather around the best-located unit. Garden forecourts were still a feature of these house-styles; terraces are often short and discontinuous, and intermixed with detached and semi-detached houses. In this type of development the estates merely leased land as and when it was required, in small lots. Examples of this type of development within the coalfield tracts are naturally small in scale, and localized in occurrence. Perhaps the most important was the settlement growth characteristic of Maesteg

---

PLATE VII. WYNDHAM, OGMORE VALLEY: AN **Ab** UNIT. This example of an adjunctive unit was built after 1900 on land belonging to the Dunraven Estate. The regularity of ground plan and simplicity of outline is characteristic, and its compactness is emphasized by the sharp break with valley floor and mountain side. The large primary school gives an indication of initial population. The pre-mining landscape is represented by Ffron-wen Farm and its fields, in a typical mid-valley-slope location. Note the playing fields on a levelled tip in the foreground.

throughout the first, and most of the second phase of settlement growth. During this period 1850–1890, the ironworks and associated collieries were rather static in terms of output, and often erratic in opening and closing with general trade conditions. Ultimately, in the mid-1880s, this iron-based economy collapsed, so that the whole period was one of greatly reduced activity. Much of the 'pre-Register' settlement in the Llynfi valley is of a mixed type, involving a considerable amount of ribbon development, and the type is shown in more detail in Fig. 14, where the irregularity of building plot size and alignment, and also the irregularity of size of the house blocks, are well shown.

In most other valleys, the dominant impression on settlement form has been to introduce patches of rather looser, less regular development within the overall settlement structure, which represent the additions to settlement during such sub-stages. Because of their easily-recognized diagnostic characteristics, they can generally be readily identified in the field. Much of the settlement in Ogmore Vale in the second phase is of this type (Fig. 9).

## THE SECOND STAGE OF THE MODEL

During this, the second development stage within the valley-unit, a number of important new factors have been introduced (Fig. 4C). The most important is that this stage of growth began between 1870–1880 approximately, and thus was broadly coincident with the introduction of by-laws for house construction, which was to have significant repercussions on the settlement pattern.

The size of the colliery units has now become progressively larger, so that the new collieries shown as opening during this stage of the model are considerably larger than their earlier counterparts. Moreover, under the impetus of a still-expanding market for coal, the railway system has been extended into the main tributary valley of the model. A new colliery has been opened far up this tributary valley, and another new colliery has been sunk in the main valley. The labour force employed at the existing pits would also show a considerable expansion as new working faces were opened. Two slants close in this phase; this could either be due to the fairly rapid 'economic exhaustion' of the mine, or perhaps to the competition for colliery labour from the bigger mines. The net result is a further upward surge in colliery employment within the model.

The associated growth of settlement has been concentrated in two locations. The most conspicuous is the large new pit head settlement alongside the new colliery in the tributary valley at A.

The large size, compactness and regularity of outline is characteristic of this type and age of settlement. The particular settlement shown has a nucleus of company-built terraces, but the bulk of the streets making up the grid-plan layout are composed of private speculative housing, with a smaller share belonging to building clubs of colliery workmen. Across the river from the main nucleus, a row of 'mixed' construction clings to a steeper, less-favoured site. The entire settlement unit would be constructed in a short period, possibly ten years or less, and is the type often classed as the 'typical' colliery settlement; yet in its uniformity and close association with one colliery, the type is not as common as one might expect.

The second area of expansion is around the major settlement in the main valley (B), where the extensions have taken the form of two large adjunctive housing developments, one to the south of the nucleus (C) and another across on the opposite valley-side (D). Both are compact, regular blocks, and owe their form to estate-controlled development. The major housing agencies involved are the speculative builders and building clubs. Each of these new blocks could possibly have a separate name, but in both cases their degree of functional dependence on the existing nucleus is considerable. It will be noticed that most of the isolated units of the initial stage still remain unchanged.

*Predominant house types*

The morphology of both types of settlement is very similar since the same agencies were involved. The terrace is predominant, but the individual houses are larger, and have wider frontages than previously. Small back gardens and rear lanes are standard, but forecourts are absent, and the houses rigidly unadorned so that frontages present a stark monotony. In this stage the addition of extensions to the rear of the houses was a common method of increasing the living space; these could be either one or two storey. some accretive growth also occurred in the main nucleus, essentially

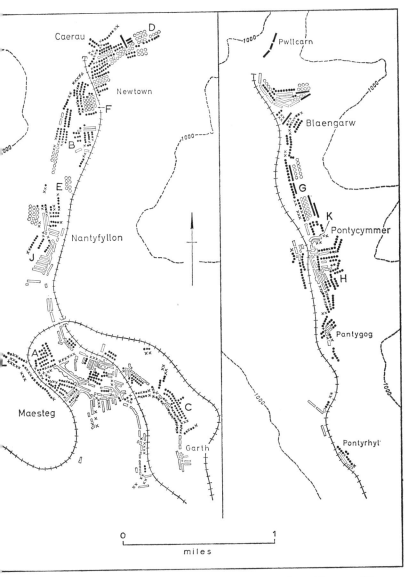

Fig. 8. Agencies of Housing Provision in the Llynfi and Garw Valleys. The letters refer to features mentioned in the text. The registers date from 1882 in the Llynfi valley, and 1887 in the Garw valley.

taking the form of ribbon development of semi- and detached houses. In the main commercial street a considerable amount of shop conversion would also be present, together with the construction of some specialized buildings, such as a workmen's institute. There is a limited amount of contract-built housing outside both main settlements—at the down-valley road junction, and along the parish road in the upper section of the valley. In the latter area, whilst employment at the collieries has increased, most of the additional labour force will be brought by contract services from the main settlement B.

*Selected case studies of new pithead settlements*

Examples to illustrate these trends within the coalfield plateau are very numerous, reflecting the very real importance of this phase of development.

Perhaps the most striking category of settlement was the large pit head unit, closely bound up with the colliery (or collieries) which it was constructed to serve.

Figures 5, 7, 9 illustrate settlement construction in the Afan, Llynfi, Garw, and Ogmore valley system during the second phase, and it will be seen that the additional settlement can be classified into that which occupied new and comparatively isolated sites, and that which had some contiguity with existing settlement.

To the latter belonged the small extensions to Glyncorrwg and the larger extensions to Cymmer, which had been a collection of cottages in 1878. Both illustrate the growth of nodal centres within the valley system, and continued the re-location trend which had begun before 1878. Apart from Gelli, there was no repetition of the creation of small, isolated settlement units, and equally significant, the existing units were not extended. The labour force for the new collieries at Argoed, Cynon, Nantybar and Cymmer Glyncorrwg was housed at Cymmer and at existing small units; considerable numbers were recruited from the lower Afan valley settlements of Cwmavon and Pontrhydyfen where the metal industries were stagnating, and these walked the three to six miles daily each way. The small expansion at Glyncorrwg was related to the uncertain position of the Glyncorrwg pit for the entire period. It can be seen, therefore, that many collieries in the valley had no associated settlement in this phase.

The former category of settlement units were still closely orientated to the collieries they served. These new units were massive in size and concentrated in plan. The 'twin' settlement of Aber and Blaengwynfi was in 1898 by far the largest settlement in the Afan valley, with the Avon pit as its focus. Abergwynfi was almost entirely constructed along rigid grid-plan lines, and Blaengwynfi was contructed in long, parallel terraces aligned along the old parish road to Blaengwynfi farm, on the steep south-facing slope of the valley. Slants were also opened nearby, which encouraged further settlement growth. On Figs. 7 and 9 Caerau and Blaengarw were the two new pit head settlements, the latter being built early in the 1880s, and the former in the 1890s so that considerable contrasts exist in the dwelling types; in fact Blaengarw bears many external resemblances to Nantymoel. It is to be noted that housing by-laws were not introduced into the Garw valley until 1887 in fact. In the Llynfi valley the widespread closures of slants, as the ironworks finally became defunct, displaced much labour. There was thus little in the way of settlement growth north of Maesteg until the sinking of the Coegnant and Caerau pits as part of a sweeping reorganization of the structure of the bankrupt Llynfi and Tondu Iron Company carried out by Colonel North after 1888.

Further very good examples of this type can be seen on Fig. 11, at Maerdy and Tylorstown in the Rhondda Fach; and elsewhere within the Coalfield Plateau, at settlements such as Llanbradach (lower Rhymney), Treharris (mid-Taff), Cilfynydd (lower Taff), Senghenydd and Ynysybwl in minor valleys. The form is a distinctive element in the pattern of settlement growth, but it is by no means as common, or as dominant as might be expected. The form of these settlements, as in the model, owed its major characteristics to a combination of the building agencies involved, and the urgency with which a large labour force had to be assembled and settled as the colliery opened. Clearly, a measure of *initial* isolation from existing colliery settlement was also a necessary factor—thus no such units are shown in the Rhondda Fawr.

*The role of the colliery company in pithead settlements*

It would be erroneous to associate these very distinctive settlements with complete 'company' domination. Two main considerations influence the interpretation of the role of colliery companies, which in detail was a complex one. As we have already seen, although there were often sound economic reasons for a colliery company to enter into the housing field, other factors had also to be considered in the total situation. Firstly, only rarely was the company a potent settlement forming agency in the sample tracts, the chief exception being Abergwynfi where the large nucleus of grid-iron terraces (A on Fig. 6), was provided by the company. This very poor-quality housing was built by the Great Western Railway from 1884 onwards in a very isolated location, after an initial row of houses had been built on a restricted site on the opposite valley-side (Glyncorrwg U.D., R.D.P. 1884–7; 1889–4). The major part of Blaengarw (Fig. 8) classed as pre-register settlement was also company-built, the registers in this case recording only the later stages of growth. These schemes, along with Gelli row (B on Fig. 6), belong in house style to the pre-register phase, but they illustrate the degree of colliery company interference necessary in the more isolated valleys of central Glamorgan. Yet only two years after the final submission of plans for Abergwynfi, North's Navigation Collieries had only to construct one small terrace group as a nucleus for the unit of Caerau (Fig. 8); here the relative proximity of settlement at Maesteg meant that other agencies were quick to follow, so that an ample supply of houses were soon under construction. This was indeed the standard role of the colliery company agency in the second and third phases; thus similar small schemes were implemented at Merthyr Vale by Cory, Nixon and Taylor, by the Dowlais Iron Company at Abercynon, by H. Taylor and Company at Tylorstown and National Colliery Company at Wattstown in the Rhondda Fach to name a few instances (Pontypridd Rural Sanitary Authority, R.D.P. 1891, Rhondda U.D., R.D.P. 1880, 1881, 1884). The outstanding fact is that these were not attempts to house all, or even a large proportion of the colliery labour force, but rather represented schemes which created confidence for other agencies; thus large company-dominated pithead units such as Abergwynfi and Blaengarw were uncommon.

The second aspect revolves around the more positive attitudes taken to the housing function by certain companies compared with the apathy of the majority. The distribution of these companies was quite random, and usually no obvious explanation can be put

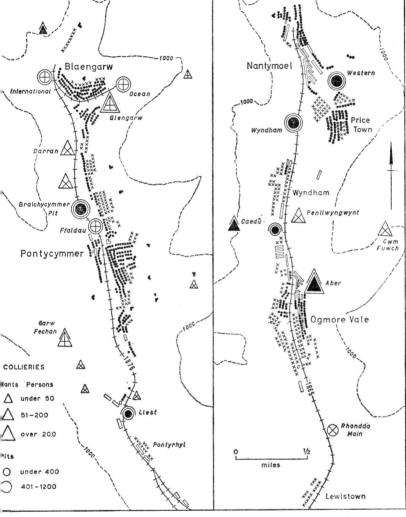

FIG. 9. GROWTH OF COLLIERY SETTLEMENT IN THE GARW AND OGMORE VALLEYS.

forward. The largest builder of company housing in the Rhondda valleys was Lewis Merthyr Consolidated Collieries Ltd., which accounted for 214 out of a Rhondda U.D. total of 583 for the post-register period—yet the focus of this company's mining activity was at Hafod in the lower Rhondda, in the heart of the densest and most continuous settlement complex in the entire coalfield where such house construction was really quite superfluous. Active company participation also took place in the Garw valley where much of the housing development on lands of Ty Meinwr farm was the work of the Ffaldau Colliery Company through its wholly-owned subsidiary, the Ffaldau Cottage Company (G on Fig. 8). Schemes submitted in 1897 and 1911 indicated a more than passing interest in property, and in 1914 the company entered the commercial property field by building a block of shops with flats over. [Ogmore and Garw, R.D.P. 1897 (497), 1911 (1253), 1914 (1468)].

However, the most important agencies involved in the construction of pithead units were the private speculators and the building clubs, and the great bulk of settlement construction was accomplished by these agencies. The property speculators, including in some valleys wealthy builders who could also finance their own construction, were the most effective bulk suppliers of housing, most capable of rapidly increasing the supply of houses in any given situation. The finance was forthcoming as soon as the likely future of the new sinking was realized, and we have many instances of large capital sums being raised on a joint stock basis. Building clubs also had access to capital through their superior mortgage status with Building Societies, and the speed of both agencies was greatly helped by close co-operation with major estates over layouts, building densities, and site works. In some situations the building clubs were also helped by a degree of paternalism on the part of the coal owners, who patronized their activities. Thus the distribution of building club housing in the Llynfi valley certainly reflects the co-operation of North's Navigation Collieries Ltd. who owned substantial amounts of land freehold in the upper Llynfi valley (Fig. 8).

---

PLATE VIII. THE UPPER OGMORE VALLEY WITH PRICE TOWN (**Ab**) AND NANTYMOEL (**Aa**). Nantymoel occupies the hillside slope in the middle distance, being a series of elongated, curving terraces. The Western colliery is still an efficient mine although sunk in the early 1870s. In the foreground is the more regular plan of Price Town, whose older second-phase section extends from the main street seen here, off the photograph to our right. The Pennant Grit plateau, recent afforestation, massive nonconformist chapels and the large spoil heap complete the landscape assemblage.

*Selected case studies: adjunctive settlement units*

The second major settlement form of this stage of the model is the growth of adjunctive units. The importance of this type of settlement development to the overall pattern is so fundamental that it transcends the division between the two later phases. Consequently the salient features of this formal element will be discussed here in terms of both second and third phase stages of development, to avoid duplication, although the actual illustrations will be restricted as closely as possible to the second phase.

The examples of adjunctive units on the Afan, Llynfi, Garw and Ogmore valleys form a useful introduction (Figs. 6, 8, 10). In these valleys, although 'infilling' of existing street outlines or the piecemeal addition of a terrace to a settlement was quite common, the dominant extensions were of a different character. They were not organic additions to an existing outline, but completely new blocks of settlement linked to the existing units only by virtue of relative proximity and through serving the same colliery or group of collieries. This type of growth is perhaps best-seen in the Ogmore and Garw valleys, where Price Town and the major part of Ponty-cymmer are good examples. The regular grid layout is noticeable, together with the sharply defined unit boundaries; the form of these units in fact suggests some type of overall plan and control. At Pontycymmer the new extension involved in effect a re-location away from the western, original nucleus, to the eastern valley-side. The new unit in proximity to the pithead settlement of Nantymoel was further differentiated through being given a distinctive name, Price Town (Plate VIII). A rather untypical feature of the phase in the Garw valley was the ribbon of terraces along the parish road joining Blaengarw and Pontycymmer.

*The Rhondda Fawr valley:* a major expansion of settlement occurred, but the employment expansion was almost entirely concentrated in the existing collieries. Consequently the already well-marked flexibility of settlement distribution continued, with much infilling in the existing major concentrations, and a further substantial amount of ribbon development along the parish roads (Fig. 11). Large adjunctive blocks of settlement were also built, notably at Ton Pentre, Gelli, Tonypandy and Clydach Vale (Fig. 12). The overall result in the Rhondda Fawr was to emphasise the maturity of the total pattern, in which ribbon development and the selected kernel-like growth of favoured nuclei were more characteristic than massive pit-head concentrations.

*The eastern coalfield valleys:* the distribution of new colliery settlement was still overwhelmingly influenced by the interaction of down-valley coal mining and head-of-valley urban settlement, which again distorted the expected colliery settlement response. The second phase was marked by the sinking of deeper and larger pits still further down valley, and including the Elliot, Pochin and Abernant collieries in Fig. 13. This ran parallel to the further run-down of employment in the ironworks themselves (some, including Plymouth, Aberdare and Nantyglo, were closed permanently), along with the closure of many small associated collieries in the vicinity of the ironworks. Increased movements of labour from the iron towns were needed if their economic position was to be safe-guarded at all, and it is apparent that in this phase there was an almost negligible amount of house construction in the valley-head settlements, and the pattern was repeated in other valleys. Only Tredegar had much new building, and this came at the end of the phase; a few streets were also added at Pontlottyn.

The major expansion of settlement occurred at New Tredegar, where mining developments were focussed, and took the form of adjunctive extensions, such as Brithdir and Elliotstown, with further streets added to other units. Nevertheless, the amount of settlement was greatly reduced by the flow of labour downwards in particular from the up-valley settlements. The degree of restriction varied according to the detailed circumstances—thus the huge Pochin and Abernant collieries in the Sirhowy were quite devoid of associated settlement. The Ebbw Vale Iron company ran services down-valley from Beaufort through Ebbw Vale to the collieries at Waunlwyd and Marine, although at the latter colliery a substantial pithead settle-ment also developed. Generally, with exceptions such as Abertillery or Treharris, or Llanbradach, the large pithead settlements so common in the tracts of the central coalfield were less abundant in the eastern coalfield, especially in the upper valley sections.

*The trend towards concentration in colliery settlement*

The increasing significance of adjunctive units in settlement growth in the coalfield plateau generally can be attributed basically

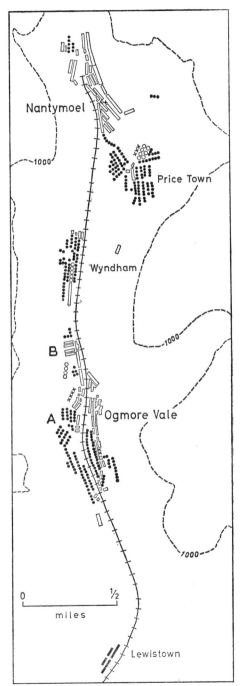

FIG. 10. AGENCIES OF HOUSING PROVISION IN THE OGMORE VALLEY. The letters refer to features mentioned in the text. The Registers date from 1887.

to two types of factors. In the first place, there was the rapidly rising pressure on building land with the massive expansion of the coal industry. Secondly, the arrival of the Housing By-laws, with a proper system of supervision under the Sanitary District system, which extended over most of the coalfield plateau by 1886.

The net effect of the increasingly high standards enforced by the by-laws was in fact to increase house construction costs, which at the same time became proportionately greater for completely new schemes, since additional expenses of sewerage, piped water and probably gas supplies would have to be incurred over and above an accretive extension to an existing settlement. In this way the attraction of the larger established centres, especially in the third phase, was not only due to social considerations but was also a reflection of economic factors. The importance of this aspect can be illustrated from the records of Llantrisant R.D., where a fragmented settlement structure characterized the rapid economic development in the third phase.

The first scheme of streets and houses to be submitted in relation to the Margaret and Mildred colliery (the present Cwm colliery) of the G.W.R. in 1910 was for the Tynynant Building Club, which failed for a number of reasons, one being simply '. . . there are no sewers provided'. (Llantrisant R.D., R.D.P. 534, 1910). Subsequent submissions by a Tynynant Land and Building Company, and the Tynynant Building Estate also failed. In 1914 a modified scheme of the Tynynant Building Club was approved. and it is significant that the application was submitted along with a G.W.R. scheme for a further 200 houses, indicating that the colliery company had assumed overall control of the situation and was prepared to offset the extra costs incurred in a large 'greenfield' housing development. Where the promoters were property speculators these extra costs were borne by the estates and the developers together, but were naturally reflected in higher lease costs and higher rentals. It is hardly surprising that many schemes were never carried out, even though approved. Also the small private builder was normally incapable of meeting all the extra costs of estate promotion, which thus tended to be the preserve of the very large speculative builders or property investment companies. Ribbon development or infilling was all that the small builder could venture. The cost mechanism in the coalfield, especially after 1880, was thus channelling most of the speculative housing into the safety of estate-sponsored and possibly subsidized schemes, and discouraging the growth of new settlement units.

*The role of the large estates in the settlement process*

In most valleys of the plateau, against a background of economic and demographic pressures, the estates played an active role in the settlement process. The large-scale housing programmes needed estate promotion and/or encouragement and co-ordination, because of the high costs involved and the necessity for ensuring a rapid output of dwellings. Consequently the majority of speculative and building club schemes, which were usually in the size range 20 to 40 houses, were situated in large and compactly planned 'blocks' of settlement.

Perhaps a key physical attribute of the large south Wales estates was the high degree of fragmentation, the estates being compromised essentially of aggregates of farms or blocks of farms; only rarely did these form extensive contiguous blocks. In no instance in the samples did one estate control an entire valley, although an estate sometimes embraced one valley-side. A further feature was the frequent random occurrence of freehold farms, which thus formed enclaves in the territory of the major estates. This fragmentation meant that no estate enjoyed a monopoly over building or mineral leases in any one area, a factor which was to be of importance for the structure of the settlement.

The better building land in any valley would obviously belong to one estate or another—yet the commencement of housing development on that land was not an automatic reflex of demand, and much would depend on the policy and commitments of the owners at the time. By way of illustration we can cite the two-phase growth of Pontycymmer (Figs. 8, 9). Development in the second phase occurred entirely on the lands of Nantyrychen farm (H), belonging to the Llanharan estate; third phase development switched in emphasis to the lands of Ty Meinwr farm (G) belonging to the Dunraven estate. There was little significant difference in either the physical or locational qualities of the two sites; the different stages of development reflected in all probability the neglect of the full development potential of the central Glamorgan valleys by the

Dunraven Estate until after 1898, a factor which may be related to the extensive interests of the latter in the richer Rhondda valleys. The same pattern is seen in the contrasting dates of development of the units of Wyndham and Price Town in the Ogmore valley (Fig. 9). It should be stressed here, however, that this is not to imply a complete absence of housing development on Dunraven lands in the second phase—a fair amount of piecemeal growth did take place—but it was unprepared, and of a low intensity. Certainly without estate initiative and controlling guidance the overall settlement structure would have been different—less coherent, probably more expensive, and less compact.

*Estate building layout proposals* were carefully prepared, their execution was a costly process, but the eventual return on capital in leasehold rentals was high. Proper provision of streets, rear access lanes, storm water drainage and sewerage, piped water and so on were all stipulated by the increasingly stringent by-laws, and the registers show that the latter were not dead-letters for even the wealthiest estates had layouts rejected on a variety of grounds. Estates in turn normally enforced minimum standards for the proposed new dwellings with a view to the capital appreciation of their estate, and most aimed at the best type of dwelling which they could reasonably expect to be constructed. The most important feature of the layouts submitted was their compact form and economy in the use of land, which facilitated the satisfaction of by-law regulations and avoided the wasteful sterilization of future building land which could easily occur with uncontrolled and shortsighted development. The grid concept was the most acceptable form for most of the period, and ribbon development was never a feature of estate schemes.

With land so prepared, even with the collaboration of the estates on essential works such as road preparation, only the larger private sector investors and stronger building clubs could afford the higher cost of the leases, but on the other hand only the larger estate-planned schemes could provide sufficient land for rapid house construction, so the benefits were to some extent mutual. Since the speculators and building clubs were by far the most important agencies in the plateau the majority of houses constructed from 1878 to 1926 found its way on to these estate developments. The sequence of this, the most important settlement process in the coalfield, can be followed in any of the registers of the plateau districts. As one example, we can quote the growth of settlement in a section of the Garw valley. An entry in 1903 reads

. . . plan of proposed new streets at Pontycymmer (Ty Meinwr) for the Dunraven Estate (Ogmore and Garw U.D., R.D.P. 1903—764).

In this and the following year successive building schemes were submitted by speculative builders and other agencies for various street blocks in the estate layout (Ogmore and Garw U.D., R.D.P. 1903, 1904). The whole process was thus fairly smooth, and generally implemented in a relatively short space of time. The houses built were usually all of a similar size and style, and also date, adding further uniformity to these settlement blocks. However, it often happened that small sections of layouts went uncompleted in the main flush of growth, and these were then filled in gradually by small 'mixed' schemes in succeeding years. As a result of private negotiations and agreements between estates and major development agencies it was not uncommon for estate layouts and actual house plans to be submitted together. Thus the Rhondda U.D. registers for 1908 include two cases in the Rhondda Fawr, at Treorchy on the Bute Estate (A) and at Treherbert (B) on the Dunraven Estate (Fig. 12). This arrangement was especially common where building clubs were involved since the estate often acted as trustee for the club scheme.

Active estate participation in the settlement process resulted in the creation of a series of compact, virtually single-phase 'blocks' of settlement which were additions to whatever settlement existed already, but at the same time possessing if sufficiently large a distinctiveness of morphology and house-type which set them apart from their neighbours. The plateau tracts of our samples were indeed dominated by this process of settlement growth after 1878.

*Examples of estate proposals and competition*

Fig. 14 is a trace of the original estate layout of 1904 for new streets in Maesteg, submitted by the Turberville Estate (C on Fig. 8). This layout, typical of streets planned and prepared in advance of house building, shows the detail of the provision of facilities under by-law regulations, and the provision of rear access and garden space. The simplicity and regularity of the outline contrasts sharply with the pre-register 'mixed' growth along the parish road, with its irregularity of alignment and small blocks of dwellings. Many layouts began in this manner—a planned estate layout which ran parallel with, and behind, the parish road which almost invariably acted as the axis of early settlement growth in each valley unit. Depending on the demand, the estate schemes then expanded outwards with further parallel streets.

Estates often found themselves in competitive situations, where much development was proceeding in a valley where more than one estate was represented, and many interesting settlement features resulted. Perhaps the most important was the overall fragmentation of total housing construction, of which the Llynfi valley (Fig. 8) provides perhaps the best illustration from the samples, since the Talbot, Coytrahen, Turberville, Dunraven, and North's Navigation estates were all intermingled in a complex manner. In the Caerau district the estates of North's and Coytrahen were interlocked; the Talbot estate controlled western Maesteg, whilst the Turberville lands were mainly in eastern Maesteg around Garth. The Dunraven estate was particularly represented around Spelter (E) and Blaencaerau (D). The result was a greater spread of house construction through the length of the valley, but this was marred by some unfortunate consequences of unregulated estate competition, including the development of some bad and isolated sites in mid-valley, cut off from the main nuclei by railway lines and rivers.

A further example of settlement fragmentation created by estate competition revolved around the Coed Ely colliery, which began sinking in 1905. The new settlement near the pit head was split between the layouts of the Tylcha Fach estate, the Garth Hall estate and Tontraethwg and Gelliseren farms of the Ynysplwm estate. The entire settlement could have been more efficiently sited at Thomastown, where an excellent physical site, with easy access to the colliery and to a railway station was being developed.

Leasehold tenure was sometimes used deliberately to foster areas of better-class housing, even though the demand for the latter was obviously rather restricted, mainly involving the rising and prosperous shopkeeper and professional elements. The Talbot estate's layout plans for their Neath Road property (L) at Maesteg (Fig. 8) is a good example, in which the specified size of the plots prohibited anything but expensive houses. A similar piece of residential 'segregation' was carried out by the Dunraven Estate at its Tyne-wydd property in the Ogmore valley (A on Fig. 10).

Finally, in connection with this stage of the model, it will be noted that in all the tracts examined, there was no repetition of the creation of small and very isolated units so typical of the pioneer stage. Furthermore, such units which developed in the first phase almost without exception did not undergo any expansion, illustrating a considerable step forward in the achievement of a distribution of settlement in socially more desirable units.

## THE THIRD STAGE OF THE MODEL

In reality, this is coincident with the third phase of settlement growth within the coalfield, in which the prosperity of the colliery labour force reached its peak with the increasing volume of coal production. It was also associated with the greatest potential for settlement flexibility, in that new sinkings were fewer in number and tended to be in remote, hitherto unexploited districts, the demand for better social conditions increased, and the network of communication by rail reached its maximum extent. Mobility of labour within the coalfield was thus enhanced by these permissive factors.

In the model valley-unit, the stage contains within it evidence of two conflicting trends in settlement distribution (Fig. 4D). The increased mobility of the labour force is given expression in the model by the construction of new stations and halts at existing settlements, new settlements and many collieries.

The new collieries opened in this phase are located some distance from the existing centres of mining, generally down-valley, although at the valley-head a new pit has been sunk adjacent to the older slant mine. The new collieries are larger and more efficient than the existing collieries.

In the main nucleus of the major valley some 'better-class' housing has been built by speculative and contract builders, both in the form of additional ribbon development, and accretive growth in the form of two quiet roads in the main nucleus; near the parish church was a favoured site. This stage is also marked by a considerable amount of building club activity in the earlier years,

channelled into compact estate developments. The latter are seen
to extend up-slope from, and parallel to, the parish road in the main
settlement concentration. There is very little expansion at the pithead settlement in the tributary valley; the limited amount of new
housing is attributable to building clubs, probably with the encouragement of the colliery proprietors. Because it lacks even a modicum
of social stratification this settlement unit does not possess even a
small 'better-class' district. In the major valley two terraces of
colliery company houses have been constructed in the main nucleus,
to act as an incentive in the attraction of scarce labour at a time
when all the collieries were expanding rapidly.

The new collieries are sufficiently removed from the existing
settlements to encourage the establishment of new pit head settlement units, besides fostering the expansion of the main settlement.
Workmen's journeys by rail from the latter, together with a large
inward flow from the down-valley urban centre (whose existence
was postulated earlier) is largely responsible for supplying the
labour demands of the new collieries A, B, and C. These services
could be either statutory or contract, or both. Nevertheless, estate
schemes for the development of housing near to these collieries,
despite the high costs and other obstacles, are an additional feature
of the settlement structure. Two schemes are shown in relation to
colliery B, both of which succeeded in attracting only a limited
amount of new housing by speculative builders. At colliery A only
building clubs have built houses, and they have also contributed to a
larger layout at colliery B. These settlement units are small and
isolated; in the model building clubs have taken the initiative, but
it could just as easily have been the colliery company, since both
were equally committed to a specific location. The majority of
speculative housing in the valley has been safely concentrated in
the main settlements, in adjunctive estate schemes. This was also in
line with the increasing social forces leading to settlement concentration. In certain situations the new pit head units could be
larger, but the collieries concerned were still essentially reliant on
'imported' labour.

The rejuvenation of an isolated settlement unit, which had
remained almost ignored since the first stage, is seen in the new
housing, mainly colliery company and building club, at the head of
the major valley, associated with the new pit.

*Predominant house types*

The house types of this stage display a greater variety than in previous stages. The terrace still remained dominant, but in the speculative schemes particularly a much greater degree of differentiation between streets was introduced (Plate V). This took the form of variations in house-size itself, of the provision of forecourts, of the introduction of bay-windows, and even on the plain-fronted terraces the quality of finish and elaboration was of a much higher order than previously. The plain-fronted terrace was still favoured by the colliery companies and building clubs, but in all cases houses were larger and more generous in their living space—three rooms and outhouses down, and three bedrooms was the norm for this stage. The houses of the 'better-class' areas were of course far larger, and were semi- or detached; in no way can they be distinguished from their counterparts in Cardiff or other coastal towns. In many speculative schemes where the land contours allowed, 'house-upon-house' types were built, adding variety to the townscape but also raising the net density of residential areas. The larger houses, and the greater degree of variety introduced into the housing at this stage, is a direct reflection of the rapid advances in real wages and living standards.

The conflict between the established settlements and the new sinkings for the attraction of housing in this stage is perhaps fundamental, since it involved a whole range of social and economic considerations. Although in general terms the force of attraction exerted by the larger, established nuclei was dominant, the position could vary between valleys, as an examination of new settlement construction in Figs. 5, 7, and 9 will illustrate.

*Selected case studies*

*The Llynfi valley:* the main areas of settlement expansion were Maesteg and Caerau, and in both cases the existing outline was extended without radical change. Large additions were made along the lines of the existing grid pattern at Caerau and Newtown. The major part of the new housing was deployed through the entire length of the valley, though at the same time being concentrated in major schemes, so that it still largely conformed to the 'controlled' settlement block pattern of the previous phase. The major new pit, the St. John's, failed to attract any new settlement.

*The Garw valley:* the only new collieries were slants, located in the upper reaches of the main valley and in the Garw fechan. The total settlement expansion was not great, but most was channelled into an adjunctive extension to Pontycymmer. There were some small isolated new developments at Pontyrhyl in the lower valley, and at Pwllcarn, near the International colliery.

*The Ogmore valley:* there was more total activity, related to the large increase in employment capacity at the Wyndham pit, two new slants, and the new Rhondda Main pit. New housing was located in three areas. The first was an adjunctive extension to Price Town; the second was the construction of additional streets at Ogmore Vale which completed a compact outline; and the third was the creation of a new adjunctive settlement unit at Wyndham, again with a regular and gridded outline (Plate VII). The net effect of these developments was to create, as in the Garw and Llynfi valleys, an almost continuous mass of settlement along the valley. However, the new settlement unit of Lewistown much further down-valley than any existing settlement represented clearly the re-establishment of the small isolated settlement unit which was prevalent in the previous phase. It was small, and strictly pithead-oriented in a phase when most settlement was mature enough to choose other types of location. Lewistown was devoid of facilities, and it must be taken as a retrogressive settlement in location and form; however, such units were more common in the Afan valley.

*The Afan valley:* new collieries in the Afan valley were widely dispersed: new pits were sunk at Ynyscorrwg, Nantewlaeth, and Duffryn Rhondda, and Cynon pit underwent a major expansion; only Ynyscorrwg pit was located near to an existing major settlement. New slants were mainly located at the head of the Corrwg fechan, or in the vicinity of Cymmer. From 1908 to 1916 the Avon pit was closed because of labour difficulties. New house construction against this background showed a remarkable range of responses. Some additions were made to the nodal settlements of Cymmer and Glyncorrwg, although at the latter the run-down of the labour force at Glyncorrwg pit parallel to the development of Ynyscorrwg tended to depress the amount of expansion. The Gwynfi unit was virtually static in this phase. But each new isolated pit had a small pithead settlement unit associated with it, repeating a trend seen at Lewistown; examples include Nantewlaeth, Cynonville and Duffryn Rhondda (Plate IX). The largest new settlement unit was Abercregan, an estate scheme whose origin is discussed in detail below.

But perhaps the most significant overall feature which Fig. 5 impresses on the geographer is the paucity of settlement in total in the third phase in this valley, even after accounting for the temporary stoppage of the Avon pit. The settlement which was established was scattered in small and functionally barren communities, and thus fragmented. Each day the collieries were supplied with labour by a vast influx from the lower Afan, Llynfi and upper Rhondda valleys. The propensity towards the growth of nucleated units, which we observed in other valleys, was lacking, although the potential in site and location at Cymmer was almost ideal—indeed, since 1945, large new housing estates have been built in this location.

*The Rhondda valley* (Fig. 11): in the third phase new collieries were opened in the upper reaches of both valleys, at Mardy, Tydraw and Glenrhondda, and only the Lady Margaret pit ceased production. The potential for settlement flexibility in this phase, already considerable, was further extended by the excellent electric tramway system which began operating in the lower Rhondda in 1905, and soon ran the length of the two valleys, as well as the major tributaries. Once more the extensions to settlement tended to consolidate the existing and more favoured nuclei, so giving the settlement mass a more organic, kernel-like growth pattern than in any other valley examined. All the nuclei in the Rhondda Fawr expanded, although most of the extensions were on a more limited scale than in the previous phase; all were added unobtrusively to the existing outline. It is worth noting that many of the more isolated units also expanded, particularly Cwmparc, Blaenycwm and Blaenrhondda. There was comparatively little development in the Rhondda Fach, but the remarkable ribbon growth between Mardy and Ferndale along the tramway is a distinctive feature.

*The eastern coalfield valleys*

The third phase was a period of resurgence in house construction in the settlements in the tract of the eastern coalfield (Fig. 13).

Much new building took place at the heads of the valleys, integrated into the existing plan. Yet the complex mineral lease structure of the Rhymney valley once again adds a note of discord. The Mclaren Merthyr colliery, a large new sinking, belonged to the Tredegar Iron Company, and a compact pithead settlement more characteristic of the second phase was quickly established at Abertysswg. Undoubtedly the new housing activity would have gravitated to the existing unit of Rhymney had the ownership position been different.

During this phase the more remote settlement units were also generally extended. Examples include Deri, Hollybush and Fochriw, and although in total these extensions were tiny they did represent significant additions to, and rejuvenations of, these small units. Furthermore, new collieries were developed still further down valley, as at Groesfaen, Markham and Oakdale. The latter collieries were associated with fairly substantial but nevertheless isolated pithead settlements which were established after 1918, and again constituted examples of the retrograde 'hiving off' process which has already been noticed in the central coalfield. The major down-valley nuclei were also extended by the addition of blocks of settlement; thus at New Tredegar a new adjunctive unit, Phillipstown, was grafted in to the complex.

*Pithead settlements: the role of estates*

So far our attention has been directed at the more developed valleys of the plateau, where there was only a slight risk of outright failure, and where the estates organised fairly efficiently a lucrative private housing sector. The new estate developments were either almost adjacent to, or quite close to, existing settlement, so minimizing the financial risks. However, in the less-developed valleys of the plateau, and in the plateau fringe areas the situation was less secure, even in the third phase, and new collieries were at greater distances from established settlements. This tendency was seen especially in the Afan and Ely valleys, where the pull of the existing small units was tested to the full by the revival of pithead units in the most isolated of locations. In the third phase housing development alongside these new collieries in any large degree was not attractive to most private builders, but this did not deter private estates from forming their own house-building subsidiaries and submitting development schemes.* The combination of land and building company was a powerful one, but even so many schemes failed to materialize. Indeed the risks were often considerable, since in many areas there was no guarantee of the success of a sinking; thus the abandonment of the large Whitworth sinking in the remote Pelena valley (O.S. Grid reference 2797 1966) resulted in the shelving of many large building projects (Neath R.D., R.D.P. 1907–681–685 incl.).

The successful schemes in the Ely valley were at Thomastown, where the Ely Valley Building Syndicate built 100 houses on the lands of the Ynysplwm estate in 1907, to be followed by many extensions (Llantrisant R.D., R.D.P. 1907–350). The Garth Hall layout was only partly built on by a building club, whilst the layout of the Tylcha Fach estate was similarly never completed in its entirety. The major speculative venture at Cwm Colliery was also shelved, to be resuscitated a few years later by the colliery company.

A second example of a successful scheme from the sample tracts is that of Abercregan (Fig. 6), a tenuous terrace over a mile in length, perched precariously on a ledge which was literally hacked out of a precipitous valley side, adding a parallel terrace where the slope ameliorated slightly, in the Cregan tributary valley (Plate II). This remarkable unit is perhaps the clearest example of the determination of an estate to develop an inherently unsuitable land holding for housing. Three large and comprehensive housing schemes for the whole estate, which stretched from the Corrwg valley to the Cregan valley, were submitted as the major new collieries of Cynon, Dyffryn Rhondda and later Nantewlaeth, were developing. The schemes of 1905 and 1907 were never proceeded with in their entirety, although one small section of the outline was built on in terrace style; in the 1907 plan there were reserved sites for a school, chapel and shops (Glyncorrwg U.D., R.D.P. 1905–18, 1907–44). The final scheme was submitted in 1910 by the Cefn y Fan estate company, concurrently with building plans for over 200 houses by the same company (Glyncorrwg U.D., R.D.P. 1910–20, 21). This scheme was fully implemented, under conditions of great physical difficulties on a site cut off by an unbridged river gorge and two railway lines from the collieries it was designed to serve. Across the

---

*The Tynynant Land and Building Club even included their own brickworks, to be built near the proposed site (Llantrisant R.D., RDP, 1910–583).

valley, extensive, gently-sloping tracts of land have remained unused until the present day; it is a timely reminder that even in the Afan valley, with probably the severest relief conditions in the central coalfield, settlement growth was guided by a set of factors which were the reverse of deterministic in a physical sense.

*Pithead settlements: the role of colliery companies and building clubs*

The isolation of many of the third phase settlements also acted as a stimulus to two other agencies to engage in the expensive development of housing in 'green-field' sites—with the isolation and expense being major contributory factors in keeping the total size of the units generally small. But, particularly with the colliery companies, the acute labour shortages which developed after 1910, paralleled by the evident inability of the private agencies to cope with the housing demand, also acted as a further stimulus. Because of the isolation factor, many colliery companies found themselves building a long way from existing settlements, since their conception of workmen's accommodation was intimately bound up with the actual pitheads. These later colliery company schemes were also limited in size, 200 houses being about the maximum in the tracts analysed. The settlement structure is thus marked by the presence of these small, remote units devoid of ancillary facilities. Nantewlaeth and Duffryn Rhondda (Fig. 6), Pwllcarn (Fig. 8), and Lewistown (Fig. 10) are examples, and all consisted essentially of green-field developments. Similar schemes in the eastern coalfield provided larger units such as Markham, Trethomas and Oakdale. In some cases the new collieries were fairly close to existing settlement, in which the company sometimes provided new dwellings adjoining the existing unit. Examples of isolated settlements being extended in this way include Glyncorrwg (Fig. 6), and Blaenycwm (Fig. 12).

Building clubs closely paralleled the colliery companies in the construction of small, isolated 'fragments' of pithead settlement in the third phase, particularly building clubs associated with individual collieries. The net result was yet another 'hiving-off' process, as seen at Cynonville and supplementary housing at Dyffryn Rhondda on (Fig. 6). In some cases the determination of the building clubs succeeded where the property speculators failed—thus the building club housing at Coed Ely was built on a small section of a

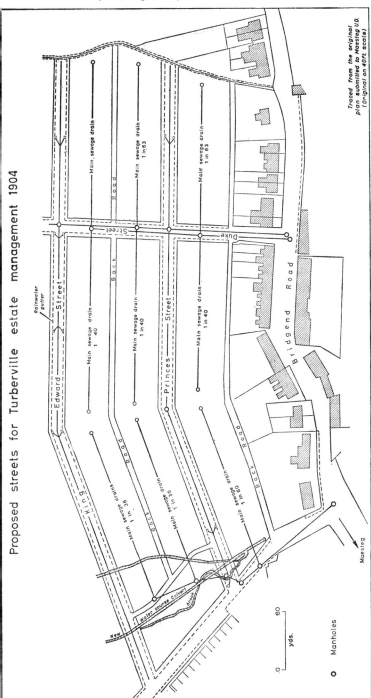

FIG. 14. AN ESTATE LAYOUT PLAN FOR HOUSING DEVELOPMENT. This figure is based on the original plan submitted to Maesteg U.D. in 1904. Existing housing along Bridgend Road is shaded.

much larger speculative scheme which was approved but not implemented in full (Llantrisant R.D., R.D.P. 363, 396–1907). At Dyffryn Rhondda the building club and colliery company submitted plans together on the same basic layout, a good combination of the committed agencies (Glyncorrwg U.D., R.D.P. 1911–18). Under the circumstances many schemes failed to materialize— that at Nantybar across the Afan valley from Dyffryn Rhondda being a good instance (Ibid. 1905–15). These two agencies were thus to a considerable extent locationally committeed in practice, whilst other agencies were not. Although the total number of houses involved in these small schemes was relatively insignificant, they have added an interesting and distinctive element to the structure of colliery settlement in the coalfield.

*Land ownership and settlement form*

Finally, in both the second and third phases we have seen that in the interaction between estates and housing agencies, the former generally played an *active* role, participating in the design and submission of housing layouts. In retrospect this unified estate control was probably instrumental in making the final form of settlement more satisfying, and the whole house construction process smoother. The present urban plan in the developed valleys serves to remind us of the importance of estate boundaries in the urban structure. The more usual, and beneficial process was the discarding of the irregular field patterns as building proceeded under the typical situation of a unified estate scheme; it was a process repeated endlessly throughout the valleys. Nevertheless it can sometimes be seen how the size and layout of streets has been closely adjusted to the shape of the previous field boundaries where there was a pronounced intermingling of estates, as in the Llynfi valley which

PLATE IX. DYFFRYN RHONDDA. A **D1** UNIT. One of many large collieries sunk after 1900 in remote locations away from existing settlements, this colliery (1908) typically received the majority of its labour supply daily by trains from further up and down the Afan valley, and the Llynfi valley. A few groups of company and building club houses comprised the original settlement, which always remained quite tiny in relation to a colliery with a peak labour force of almost 1,800 men in 1920. It reminds us that scale of settlement and size of mine is one of the most complex relationships within the coalfield.

contains many examples. In the larger valleys, therefore, where many estates were represented, this was an important cause of detailed variations in the urban 'grain'.

From the model it is clear the mining settlement in the coalfield plateau did not proceed in a haphazard and *ad hoc* manner, but responded to certain dominant influences, and thus took on recognizable spatial characteristics. Using the model we can arrive at an outline genetic classification which will help us understand the structure-patterns of colliery settlement. Some of the sub-divisions have not appeared in the model but will be readily recognized from previous sections.

CHAPTER 4

AN OUTLINE CLASSIFICATION OF COLLIERY SETTLEMENT
AND ITS RELATIONSHIP TO SERVICE AND SOCIAL FACILITIES

THIS chapter really has two objectives—to conclude the analysis of colliery settlement growth by presenting a classification of the total settlement distribution in the model; and also to suggest, by application to real examples, some important links which exist between the classification and aspects of the present social geography.

## A CLASSIFICATION OF COLLIERY SETTLEMENT

The classification of colliery settlement in the model is shown in Fig. 4E. The basis of the classification in the coalfield plateau is the *valley-unit*, which includes any tributary valleys. It cannot be over-emphasised that the methods of analysis, and the system of classi-fication, have been arrived at with the valley-unit as the common denominator and the spatial framework within which settlement characteristics were established. Within the valley-unit all colliery settlement was interwoven into an interdependent functional complex which also embraced collieries, and transportation. Consequently *all* of the following sub-divisions of the classification fall within the larger valley-unit:

**A.** *Composite Colliery Settlement*

This is the most important form, and is indicated in Fig. 4E by **Aa**, **Ab** etc. Perhaps the most important feature of composite settlements is their large size, which can often embrace the major part of a valley-unit. The Ogmore and Garw valleys, or on a grander scale the Rhondda Fawr from Hafod to Treherbert, are examples in which the composite colliery settlement is almost coterminous with the valley-unit.

The essential components of a composite colliery settlement include:

**Aa:** a nucleus of early, first-phase development as seen in the initial stage of the model. This would often include colliery company housing. It normally contains the present commercial centre, but not inevitably so.

**Ab:** large, compact and regular blocks of settlement, constructed in the second and third phases, bearing evidence of estate control. Some, or all, of these blocks might possess individual names.

**Ac:** a considerable amount of accretive expansion in all phases, containing many variations in house-type and particularly characteristic of periods of slower growth.

Some Examples:

(i) the lower Cynon Valley (commonly called Mountain Ash with adjacent settlements);

(ii) New Tredegar complex;

(iii) the major sections of both Garw and Ogmore valleys;

(iv) Abertillery.

*Composite colliery settlement—an example:* Fig. 15 illustrates in some detail the classification applied to the Ogmore valley composite colliery settlement, itself part of the Ogmore Fawr Valley-unit. The boundaries of the various units are emphasized by the cartographic technique used, and quite a complex pattern in fact results. Perhaps the main feature is that two nuclear units are present—at Nantymoel, and in the south at Ogmore Vale—the latter extending on both sides of the river and main valley railway. The Nantymoel nucleus, at the head of the valley, began as an early pithead unit and its cramped, terraced site has witnessed only a limited amount of accretive development at either extremity. By way of contrast the better-located Ogmore Vale nucleus has expanded considerably in an accretive manner, particularly in a southerly direction along what was then the main valley (or parish) road dating from rural origins.

Small fragments of settlement from the pioneer stage have still not been incorporated completely into the built-up area, as at Aber Houses; this retains a **C1** classification and is thus strictly speaking not in the composite unit.

The greater part of settlement present consists of settlement in adjunctive blocks (**Ab**). Each has a characteristic 'grid grain', and similar date of development—thus Price Town proper is second phase; the adjacent units third phase, and the greater part of Wyndham also third phase (although a pre-1878 row is included in this unit). the units are quite sharply differentiated with the exception of the divisions in the south of the settlement. There is little that can be described as 'ribbon' development in the valley apart from the Ac extensions southwards from the Aa unit at Ogmore Vale, where some irregularity of block size, frontage line and house style occurs. Similarly the addition of terraces above this parish road axis at various dates falls into this category; otherwise the channelled and phased nature of settlement is clearly dominant. With the exception of the early Nantymoel—Western Colliery relationship, there is no strict pithead dominance. Mines have been well distributed in the valley, and also slants on the upper mountain-side and in tributaries such as Cwm Fuwch. Finally, in terms of the structural plan, we might comment on the considerable number of farms which have survived physically and functionally from the rural pre-1860 period, such as at Talga, Aber, Ffronwen and Cyffog. This is quite a common state of affairs in the plateau valleys because of the mid-valleyside siting of the majority of farms, so placing them well above the level reached by industrial settlement.

This classificatory system of colliery settlement based on form and genesis is of considerable significance for the fuller appreciation of many problems which arise in the field of urban planning at the present time. Consequently, and mainly within the context of the examples being outlined, the relationship of cultural and service provisions with the present classification is commented upon. The data was collected in the field in 1962–1963, and is intended as supplementary evidence. It should therefore be borne in mind that the approach is only at an elementary level, and it is not intended that this should in any way take the form (or place) of a detailed analysis of functional organisation patterns within the coalfield, in which field W. K. D. Davies has already made a powerful beginning (Davies).*

In terms of the distribution of retail facilities, which are taken as diagnostic of the most important service provisions, there are within the composite settlement of the Ogmore Valley four con-

---

*Furthermore, the Glamorgan County Planning Department have published a valuable study on the overall social and economic problems within Glamorgan. (Powell, 1965).

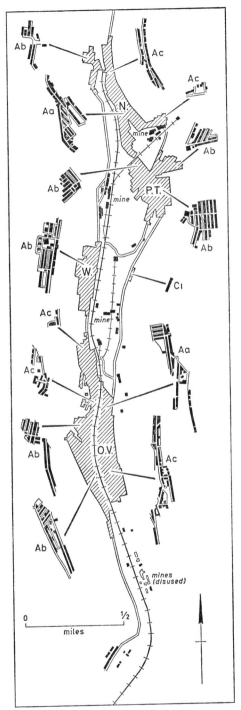

Fig. 15. Application of the Settlement Classification to the Ogmore Valley Unit. The *actual* extent of the built-up area is shaded. N—Nantymoel; W—Wyndham; P.T.—Price Town; O.V.—Ogmore Vale.

centrations, whose location is illustrated in generalized form on Figure 15. In rank order of number of shops, the hierarchy is:

(1)   Ogmore Vale   (**Aa**)   (45)
(2)   Nantymoel     (**Aa**)   (34)
(3)   Price Town    (**Ab**)   (21)
(4)   Wyndham       (**Ab**)   ( 9)

It will be seen that the nuclear units (**Aa**) still dominate the local service structure by a considerable margin, and that the more accessible (in terms of the valley population distribution) Ogmore Vale has achieved dominance. However, with the possible exception of Ogmore Vale, all the centres are rather poor. Thus Wyndham caters for very localized needs, possessing a basic range of food shops and general 'convenience' shops such as newsagent and tobacconist. Price Town and Nantymoel, in older and larger units, are also larger in their total shopping provision, but they still possess little more than neighbourhood services. Neither of the lower three has a bank: but Nantymoel does possess a co-operative 'Emporium' as indicative of slightly higher status. All the three units possess infants/junior schools, but the 'initial advantage' of the nuclear unit of Nantymoel only really becomes apparent with social facilities, especially places of worship: thus Nantymoel has six chapels, Price Town three, and Wyndham merely one.

The better-located Ogmore Vale unit (**Aa**) is the only unit to which we can truly ascribe anything other than neighbourhood functional services. Within its larger total of shops it includes a co-operative emporium, a Wales Gas Board Showroom, two Banks, a cinema, and branches of 'multiple' chains such as Peglers. This refers to the position in 1962–63; more recent fieldwork in Easter 1968 has revealed evidence of a considerable decline in social provisions since 1962–63, particularly in Chapels and public houses, and the closure or conversion of cinema premises. It clearly plays the role of a local sub-centre for its composite colliery settlement but little more. Socially it is also more important, containing nine chapels, a feature of great significance in a sectarian nonconformist population. But the limited range of 'urban facilities' in a more complete sense (newspapers, administrative, etc.) together with the relative poverty of its retail services place it well down in any wider hierarchical structure, and it would fail any standard testing method devised by urban geographers. Yet, these types of social and service patterns are very representative of their classificatory type in the coalfield plateau, highlighting the social inadequacies of contemporary settlement in the valleys. The equivalents of Ogmore Vale, Price Town or Nantymoel are all too common amongst Aa and Ab units of the coalfield, and the lowly status in any wider perspective of Aa units such as Ogmore Vale reflects the enormous problem of providing a really urban infrastructure in the coalfield. At this point, some of the findings of W. K. D. Davies are of immediate interest. Davies, in his analysis of central places in the Rhondda valleys system has established, on the basis of computed 'centrality values'* for forty-nine separate types of retailing outlets, a grouping of central places into five distinctive classes and fifty-one central places were identified'. One of the most significant facts to emerge in relation to our particular problem is the tremendous quantitative gaps separating the members of the various classes. Thus, using the ranking put forward in Table II, based on the summed centrality values for each central place which forms the 'Functional Index'; and using the classes established by Davies, we can arrive at the following general pattern:

| Class | Range of Functional Index (integer values) | Number of Central Places |
|---|---|---|
| A | 2078 | 1 |
| B | 250–451 | 3 |
| C | 84–176 | 4 |
| D | 12–59 | 18 |
| E | 2–12 | 25 |

This table highlights the disparity between the lower D and E classes, which are numerically more important as one would expect, and the remaining central places in this hierarchy. Characteristically, the leading eight central places occupy **Aa** units with the added advantages of a strategic location within the essentially linear mass of settlement. All other types of units under our structural classification fall into classes **D** and **E**. Thus **Ab** units such as

---

*Basically developed from the coefficients of Localization used in industrial geography; each type of Retail Outlet is considered in terms of its overall distribution within the system—some types of outlet being almost ubiquitous, others highly localized or concentrated.

Gelli, Williamstown, Ton Pentre and Tylorstown are in this category, similarly **B** units such as Ynysybwl, Cilfynydd and Maerdy. It should also be noted, however, that the early growth of accretive ribbon settlement in the Rhondda Fawr has given **Aa** status to some units which fall into the lower classes of Davies' functional hierarchy, such as Ystrad, Cymmer, Tynewydd and Trehafod.

**B.** *Second-Phase Pithead Settlement Units*

This class of colliery settlement is smaller than **A**, but nevertheless often had a population in excess of 5,000 in the peak of mining activity. A compact outline, a basically gridded layout, and supremely monotonous single-phase terraced housing are characteristic formal elements. Initially, at least, they were always orientated distinctly to one colliery, or well-identified group of collieries. Examples*:

| | | | |
|---|---|---|---|
| Maerdy | 297198 | Cilfynydd | 308192 |
| Treharris | 309196 | Senghenydd | 311190 |
| Ynysybwl | 306194 | Cwm | 318205 |
| Llanbradach | 314190 | Blaen-Aber-gwynfi | 289196 |
| Abercynon | 308194 | | |

*An Example—Gwynfi, Afan Valley:* the 'twin' settlement of Blaengwynfi and Abergwynfi is smaller than examples such as Ynysybwl or Llanbradach, so that its range of services and other provisions is probably poorer than the average for its class. Thus in 1921 the settlement had a population of 4,124, compared with 5,133 at Ynysybwl; whilst the rapidity of subsequent population decline to a mere 2,659 in 1961 has undoubtedly emphasized its present functional poverty. However, in its essentially mono-phase construction and its clear orientation to one group of collieries, it does present the essence of this classificatory type. (For details of development see chapter 3).

The settlement had a mere twenty shops in 1963, including a co-operative 'emporium'; these serve only local needs for convenience goods; although the population is large enough to support specialist electrical goods retailers, for example. There is one bank,

---

*For these, and succeeding examples, 6-figure National Grid references based on Ordnance Survey 1 : 63360 sheets 153,154 will be quoted for identification purposes.

which functions on a part-time basis only. Other social provisions are marked by a Workmen's Institute, four chapels and one church. Because of their larger size, most **B**-class units in the coalfield would rank higher in service provision, possessing a similar range to that found at Ogmore Vale, whereas Gwynfi clearly is functionally approximate to Price Town unit.

**C.** *First Phase Pit head or Slant Settlement*

These settlement units, originating in the first phase during the pioneer colonization of a valley-unit were closely orientated to erstwhile slants or pits. They have remained small, and slant settlements in particular are often in isolated locations. We can recognize four sub-types, of which only two were represented in the model.

**C1:** larger settlements orientated towards first phase pit heads, and which have remained relatively static; the sub-type is rare, since most became the focus of further expansion and so are included in class A. Examples include Abercanaid and Troedyrhiw in the Taff valley. (305203 and 307202).

**C2:** small, isolated terraces or cottage groupings, often row type, which form fragments of colliery settlement. Amongst many examples are Cwmcas (288199), Nantybar (284195), Sebastopol (313204), Fernhill (292200), Penybanc (311203), Fforchdwm Cottages (281197). Some have now been abandoned.

**C3:** as C2, except that some rejuvenation of house construction took place in the third phase. Examples include Deri (312201); Pontyrhyl (290189); and Glyncorrwg (287199).

**C4:** as C2, except that their distribution is confined to zones with ironworks and former ironstone mining, so that they were multi-functional, housing coalminers, ironstone miners and ironworkers. This sub-type is particularly developed in the section of the northern outcrop between the upper Cynon valley and Tredegar. Examples include Pantywaun (309207); Clwyd-y-Fagwr (302206).

*An Example of* **C3** *settlement:* Pontyrhyl is a collection of rows in the low lands of the Garw valley and extending into its tributary, the Rhyl. This settlement originally served a considerable number of small slants in the vicinity, including some high up in the Rhyl valley. By-passed by the development of deep shaft-mining in the upper Garw valley after 1876, the settlement was rejuvenated partly by the re-opening and enlargement of the local slants after 1900. New building mainly occurred along the new down valley road after 1900. The 'rural' level, represented by farms such as Lluest and Ffawyddog, remains undisturbed.

**C3** settlement units (of which other examples such as Glyncorrwg or Blaenrhondda are larger) have a limited range of social and service provisions, but the social facilities owe much to the 'initial advantage' of early establishment possessed by these units. Thus Pontyrhyl has only five shops, all of the provisions type and catering for daily needs. Yet there are three places of worship, a school and two inns, a fair total for such a settlement, and providing a 'village-type' of infrastructure.

**D.** *Third Phase Pit head Settlement*
The size-range can be considerable, but they never approached those of the second phase (Type **B**) in terms of population. They are characterized by a pronounced disparity between population and employment afforded by the new collieries to which they were related. Generally they were built with strong contributions from locationally-committed agencies, such as colliery companies; three sub-types can be recognized:

**D1:** diminutive, isolated fragments.
Examples include Dyffryn Rhondda (284195), Cynonville (282195), Lewistown (293188), Groesfaen (313200), Tynynant (306185), Nantewlaeth (286197), Wyllie (317194).

**D2:** larger settlements, fostered by estates.
Examples include Thomastown (300186), Abercregan (284196).

**D3:** 'cité minière' type, constructed by colliery companies.
Essentially a feature of the post-1919 period in the Blackwood Basin, such as Markham (316201), Oakdale (318198), Trethomas (318188), Taff Merthyr Garden Village (310198).

*Class **D1**—some examples:*
Dyffryn Rhondda: colliery company housing of 1912 supplemented by local authority and building clubs before 1926.

Dyffryn Pit began sinking in 1908, but a high level slant existed in the 1890s. Very elementary facilities consisting of four 'provisions' shops, a junior/infants school, and a workmen's Institute. No places of worship.

Cynonville: Building Club housing immediately pre-1914. No facilities of any description; associated with Cynon Pit.

Nantewlaeth: colliery company housing, with no social facilities, built in 1913. Associated with the Nantewlaeth colliery (1913).

Lewistown: 1914, colliery company housing for Rhondda Main Colliery, 1910. (Lewis Merthyr Consolidated Collieries Ltd., hence Lewis-town). Two shops on main valley road constituted its services.

In the case of Lewistown, considerable inter and postwar municipal housing developments have taken place, so that the attribute of isolation is now breaking down. With the many examples of this class in the coalfield of varying sizes in population terms the service and social provisions will differ slightly, but not fundamentally. At worst, their small size discouraged anything in the way of community facilities; at best, they have a small number of shops, possibly a school and institute, but (a sign of social changes already running strongly by 1910) rarely places of worship.

*Class* **D2** —*an example:* the distinction between the sub-types D1 and D2 is an arbitrary and loosely-defined one of *size*, but nevertheless it is true that the coalfield plateau has few examples of large pit-head units in the third phase which stand comparison with the monolithic creations of the second phase. Even the 'cités' of the Blackwood Basin are generally small in comparison with the experience of English coalfields, although Trethomas and Oakdale are exceptions.

*Abercregan:* functionally, the larger size of the D2 units ensures the presence of elementary neighbourhood needs, but little more is present. Thus Abercregan has merely six shops, all of the 'provision' type. It possesses one workmen's institute and one public house, and a junior/infants school. Despite the allocation of a site on the estate layout, no church or chapel was ever built. In 1968 the entire unit was in the throes of abandonment, and rapidly becoming ruinous through a process of voluntary rehousing to new housing estates at Cymmer.

**E.** *Expanded Villages or Hamlets*

Most of the settlement expansion was accretive, and added on to rural, agricultural hamlets. The ribbon growth of detached and semi-detached houses was common, giving a *strassendorf* form. This type is most commonly represented in the Blackwood Basin and southern outcrop zone.

Examples: Blackwood (317197); Fleur de Lis (315196); Cwmdows (320196); Tonyrefail (301188); Nelson-Llancaiach (311196); Llantwit Fardre (307184).

**F.** *Larger urban settlements based originally on ironworks*

Their chief importance derived from their large size and established urban character and facilities in 1850, at the outset of mining colonization. Many had become 'pure' coalmining settlements in functional terms by the third phase, during which a rejuvenation of house construction was common. They were immensely important as sources of colliery labour.

Examples: the urban settlements of the northern outcrop are the classic cases, particularly Aberdare, Merthyr, Tredegar and Ebbw Vale. Within the coalfield plateau a number of settlements were related to activities connected with the iron industry, such as Abercarn, or Abersychan, but particularly Maesteg.

*An example—Maesteg:* Maesteg in the upper Llynfi valley was dominated by the integrated iron and coalmining activities which focussed on the important Llynfi ironworks until the mid-1880s, when the Llynfi & Tondu Iron and Coal Company went into liquidation (for details of development, see chapter 3).

In terms of its functional facilities, Maesteg is separated by a tremendous quantitative gap from all the other classes of settlement examined, which is due essentially to its initial advantages and momentum established before 1880.

The town contains a well-defined 'central area' for retailing, and also incorporates other social provisions, such as the Town Hall which acts as a place of assembly and entertainment. A focus of retail activity is the small covered permanent market, supplemented by a large open-air market each Friday. In 1962–3 the core area contains, besides a large representation of basic 'provisions' types of shops, three variety chain stores (Woolworths, Macowards, Leslies), four major furniture retailers, a large co-operative complex, three cinemas and six banks. 'Multiples' are well represented, as are specialist outlets such as photographic suppliers, sports goods and jewellers. In the Glamorgan Planning Department's recent analysis of the thirty-one major central places of Glamorgan, Maesteg ranked thirteenth, slightly above other important service centres in the coalfield such as Tonypandy, Porth, Ferndale and Bargoed, which is indicative of its status; nevertheless all remained well beneath peripheral centres such as Neath or Bridgend (Bland).

**G.** *Urban centres with an important Labour supply function*

Although most were small, many were essentially dependent upon service functions and transport; some had other types of industrial activity, such as tinplate. Mining played a varying but significant role in the economic structure of all these settlements.

Examples: Pontypool, Caerphilly, Pontypridd, Neath.

## THE RELATIONSHIP WITH SERVICE
## AND SOCIAL FACILITIES

The pattern of settlement structure in the coalfield, which we have been studying, is significant for a fuller understanding of the problems which arise in the field of urban planning. In particular, it has been shown, for the cases examined in detail at least, that the social service and general 'community' structure which many investigations have placed considerable stress upon (see the Alcan '*Scanner's*' appraisal of Rhondda, 1966) is bound up with the physical structure pattern, itself a product of a particular type of settlement genesis. The various structure-elements have therefore important roles in the physical aspects of social geography, since they provide the framework for 'community', and also the framework for modern, late twentieth century urban living. Without denying the importance of fundamental economic changes now taking place, it is apparent that the *social* aspects of valley-living are of equal significance in the problematic future of the coalfield districts, a view recently and strongly reiterated by the Welsh Office in its appraisal of Welsh planning problems:

The problems of planning for change are most prominently posed in the mining valleys, where much of the original economic base has gone and has only been partially replaced by new industry . . . some continuity of emigration from the valleys is inevitable

*and the future levels of population and the standards of housing and
amenities in the valleys will continue to need detailed study.*

Later, the Report stresses that, for the valleys of the plateau,

a comprehensive programme of internal and environmental
development

will be necessary (H.M.S.O., 1967).

The sample tracts of our analysis display a very wide size range
from the smallest 'fragment' to the complete urban centre of
Maesteg, and all offer different conditions for contemporary urban
living in terms of residential and social provisions, as well as
possessing different potentials for development or at least adaptation.
Many apparent contradictions are present. The 'better' settlements
from the point of view of general social and service facilities are
usually the *oldest* and hence possess the largest proportion of old,
inadequate dwellings, whereas the newer settlement in many
valleys has often been constructed in bad sites remaining at that
date, and are amongst the first to be demolished—as seen at Abercregan and a section of Glyncorrwg, giving the housing in both
areas a *completed* life span of barely more than fifty years, which is
short by any standards!

It is clear that from the point of view of an understanding of
the relationships between the 'physical' aspects of contemporary
social geography and the morphogenetic classification suggested in
the paper, there appear to be five types of situations present—

(*a*) *Units* which are virtually or entirely devoid of any tangible
social provisions. These all fall into the C and D categories of the
morphogenetic classification, and many have been recently
affected by demolition. Their usefulness is clearly outlived.

(*b*) *Units* with perfunctory and elementary facilities. The unsatisfactory position of many is alleviated (and partly caused by) the
presence of a larger settlement, and there is a considerable
gradation within them in terms of provisions. Quite well-
balanced communities at a 'village' level exist—such as Pontyrhyl; but for modern needs all must be well connected with
higher-order services. The rapid run-down of such units in a
planned programme would *not* involve the wastage of much
'social capital'.

(*c*) *Units* with a complete range of 'neighbourhood services' are
dominant in the coalfield plateau. The range of retail outlets

rises above the elementary 'food-base' types of the former group, to include chemists, clothing, ironmongery, electrical goods and so on. Yet deficiencies are all too obvious to the uncommitted observer—the rarity of good multiple outlets, the rarity of banks, except on the odd, part-time basis. Social facilities are also narrowly-based upon club, pubs and chapel, with the two latter in active physical retreat throughout the coalfield.

(*d*) *the Local Sub-Centres*—which almost invariably include the original Aa nucleus (i.e. in very large valleys) of all composite colliery settlements or the dominant Aa nucleus in units where more than one was present. Possessing all the facilities of the previous group, their enhanced status is expressed eventually in the presence of banks, some multiple chain outlets, and features such as Gas and/or Electricity showrooms; also the presence (or former presence) of cinemas, and sometimes a Grammar School. Their status, however, is clearly *sub*-urban, and only the better-located and/or larger have achieved a range of facilities which is capable of earning an 'urban' status.

(*e*) *Urban Centres*—which are relatively rare in the coalfield. It must be emphasized that the quantitative gap between the local sub-centres and the centres of urban status in terms of service provisions is immense, and forms one of the most disturbing and damaging aspects of the settlement geography of the coalfield. Thus of the thirty-one centres identified and analysed by the Glamorgan Shopping Study only eight were in origin coalmining settlements, and these occupied lowly places in the ranking scheme, the highest ranked being Tonypandy at fourteenth.

The inequality of service distribution, and the barrenness in terms of tangible investments in social capital, of much of the mining settlement in the coalfield detracts from their ability to hold population in the changed social conditions of South Wales, or indeed England and Wales as a whole. It is possible for entire Urban Districts, such as Ogmore and Garw Urban District, to have no centre providing more than poor 'sub-centre' types of services; in extreme cases, such as Glyncorrwg Urban District, no unit possesses more than 'neighbourhood-types' of services. The physical structure of settlement appears singularly inappropriate in much of the coalfield for adaptation to modern patterns of living, and there are in particular two problems inherent in this situation which have a direct significance for urban planning in the coalfield.

In the first place, the phrase 'mining district' needs to be heavily qualified because it can include tremendous disparities in settlement structure, as expressed in terms of size, disposition and residential quality. Because of the great gaps in the functional hierarchy between the lower grades and the only true 'urban' centres, most mining settlements have for long been dependent on other centres for their urban services. Most of these are polarized in the urban centres of the coalfield itself (where many, especially those on the north crop, have eccentric locations), or in selected peripheral 'valley-mouth' centres such as Bridgend. In line with well-developed trends towards further specialization and concentration of retail and other services since 1950 it is apparent that, drifts of population apart, further declines in status and in total service population will occur in *all units* in the classificatory system, except the urban category.

It is difficult to see how this trend can be halted, unless there are radical changes in population movements in the region. The trends work selectively in favour of off-coalfield centres such as Bridgend and Neath, and the larger centres in the coalfield such as Aberdare, Maesteg or Pontypridd. The usual evidence of a long and continuous contraction of service provisions in the lower class units is plain to see in the field. It must surely strengthen the hand of those who would favour a selective 'pruning' of the settlement units of the coalfield in order to concentrate efforts on those with the best *urban* potential.

Secondly, the problem is obviously most acute in the units with little or no service provision. Sometimes because of age and isolation, many have already been abandoned, and their inhabitants re-housed elsewhere, so that at last a retreat of colliery settlement is under way in most parts of the coalfield from these inhospitable units. There has been no real opposition to this level of settlement relocation, but the nettle which has yet to be grasped is the future of other, marginally larger units, such as those of D1 or C3 stus.

## Concluding remarks

The object of this paper has been to throw light upon the form, structure and disposition of colliery settlement in south Wales. It is recognized that the work has been of an exploratory nature, seeking to define fields of enquiry and methods of approach rather than to provide a set of answers. It is hoped that some of the methods will lend themselves to the study of other coalfield areas, or different types of industrial settlement. In particular, opportunity exists in the field of industrial settlement study for more detailed recon-structions of the growth of settlement in all its aspects than has been attempted here. The field evidence and the documentary data are still largely intact.

Finally, it is hoped that some, at least, of the distinctive geo-graphical personalities of the South Wales coalfield will have been conveyed—the coalfield has suffered too long from its stereotyped images.

## REFERENCES

1. *Secondary Sources*

ALCAN, 1966. *A Town called Alcan—The Scanner* (with section on the Rhondda Valley). A series of broadsheets combined in *New Society*, July 28th.

BAULIG, H. 1936. *Amérique septentrionale.*
BLAND, E. A. 1966. *Glamorgan: a shopping study.*

CARTER, H. 1965. *The Towns of Wales.*
COMMISSION OF INQUIRY INTO INDUSTRIAL UNREST. 1917. *Report for Number Seven Division (South Wales).*

DAVIES, W. K. D. 1967. Centrality and the central place hierarchy. *Urban Studies*, Vol. 4, 61–79.
DAYSH, G. H. J. et al. 1953. *West Durham, a study of a problem area in North East England.*
DEMANGEON, A. 1933. *Belgique, Pays Bas, Luxembourg.*

GENDARME, R. 1954. *La région du Nord, essai d'analyse économique.*
GLAMORGAN. C.C. 1955. *County Development Plan Report.*
GOTTMAN, J. 1955. *Virginia at Mid-Century.*

HARE, A. E. 1940. *The Anthracite Coal Industry of the Swansea District.*
H.M.S.O. 1967. *Wales—the Way Ahead (Cmd 3334).*

HOWE, G. M. 1957. in Bowen, E. G. (Ed.). *Wales, a physical, historical and regional geography.*

JEVONS, H. S. 1920. *The British Coal Trade.*
JONES, P. N. 1965. *Some Aspects of the Population and Settlement Geography of South Wales*, Ph.D. Thesis, The University of Birmingham, unpublished.
JONES, P. N. Workmen's trains in the South Wales Coalfield, *Transport History* (in the press).

LEFEVRE, M. 1926. *L'Habital rural en Belgique.*
LEWIS, E. D. 1959. *The Rhondda Valleys.*

MARQUAND, H. (Ed.). 1937. *Second Industrial Survey of South Wales.*
MINISTRY OF HEALTH 1920. *Report of the South Wales Regional Survey.*
MORRIS, J. H. & WILLIAMS, L. J. 1958. *The South Wales Coal Industry 1841–1875.*
MURPHY, R. E. 1933. A southern West Virginian mining community. *Economic Geography*, Vol. 9, 152.

NORTH, F. J. 1931. *Coal and the Coalfields in Wales.*

POWELL, E. J. 1965. *Glamorgan: a Planning Study.*

RICHARDS, J. H. 1956. *Fluctuations in house building in the South Wales Coalfield* (Univ. Wales unpub. M.A.).

RICHARDS, J. H. & LEWIS, J. P. 1956. House building in the South Wales Coalfield, 1851–1911. *Manchester School*, Vol. XXIV, 189–201.

SAUL, D. 1962. House Building in England, 1890–1914. *Econ. Hist. Rev.* XV, 119–137.

SCHWARZ, G. 1959. *Allgemeine Siedlungsgeographie.*

SELECT COMMITTEE ON COAL. 1873. *Report.*

SELECT COMMITTEE ON WORKMEN'S TRAINS. 1900. *Report.*

SMAILES, A. E. 1938. Population changes in the colliery districts of Northumberland and Durham. *Geog. Journal*, 220–32.

SORRE, M. 1952. *Les Fondements de Géographie Humaine.*

THOMAS, B. 1962. *The Welsh Economy, studies in expansion*

### 2. *Primary Sources*

NLW National Library of Wales:
    Dunraven Documents
    Llandinam Documents

GLAMORGAN C.R.O. Glamorgan County Records Office:
    D/DBJ Blandy Jenkins Collection
    D/DG Dowlais Iron Company Collection
    D/DRa Randall Collection
    D/DVau Vaughan Collection

RDP Registers of Deposited Plans and Associated material
    Glyncorrwg Urban District
    Llantrisant Rural District
    Maesteg Urban District
    Neath Rural District
    Ogmore & Garw Urban District
    Pontypridd Rural Sanitary District
    Rhondda Urban District